THE INSIDE TRACT

Also by Julie D. Rubin

The Only Security Blanket You'll Ever Need
(with Sydney Langer, M.D.)

THE INSIDE TRACT

Understanding and Preventing Digestive Disorders

Myron D. Goldberg, M.D.
and Julie D. Rubin

An AARP Book
published by

American Association of Retired Persons,
Washington, D.C.
Scott, Foresman and Company
Lifelong Learning Division
Glenview, Illinois

Authorized adaptation from the edition published by Beaufort Books, Inc.

Library of Congress Cataloging-in-Publication Data

Goldberg, Myron D.
The inside tract.
Includes index.
1. Gastrointestinal system—Diseases—Popular works.

I. Rubin, Julie D. II. Title.
RC806.G64 1986 616.3 85-31781

ISBN 0-673-24840-2

Originally published in 1982 under the title *The Inside Tract: The Complete Guide to Digestive Disorders*.

A health book developed with Biomedical Information Corporation, Inc. (BMI)

123456-MAL-91 90 89 88 87 86

To my parents and to my students, who continue to teach me.
Myron D. Goldberg, M.D.

And special thanks to James Seligmann from us both.

AARP Books is an educational and public service project of the American Association of Retired Persons, which, with a membership of over 21 million, is the largest association of persons fifty and over in the world today. Founded in 1958, AARP provides older Americans with a wide range of membership programs and services, including legislative representation at both federal and state levels. For further information about additional association activities, write to AARP, 1909 K Street, N.W., Washington, D.C. 20049.

CONTENTS

INTRODUCTION

Early in my training as a gastroenterologist, I realized the necessity of a reliable and easy-to-understand book on the digestive system. Most people know little about this part of their bodies and aren't even aware how vital this knowledge can be to their overall well-being. When they're feeling well they take their health for granted, assuming things will continue this way, no matter how much they abuse their system. But when a problem arises, they suddenly become aware of how little they know and how important this information can be. They may turn to a friend or relative for advice, attempt to medicate themselves—doing more harm than good—or try to ignore the problem, hoping it will go away.

All too often, lack of knowledge causes needless suffering and anxiety. Many serious digestive disorders can actually be prevented if an individual understands some basic dos and don'ts about diet, exercise, posture, fluid intake, sexual practices, and self-treatment. Such an individual knows how to keep his or her body healthy, how to handle effectively many minor problems, and how to judge if a problem is serious enough to bring to a doctor's attention. Frequently, people are able to save their own lives through early detection of disease.

Everyone has a digestive problem at some time. It may only be an occasional bout of diarrhea, a mild stomach upset, or a case of food poisoning. On the other hand, it may be something far more serious, requiring the knowledge of a physician specially trained in the field of gastroenterology. Whatever the problem, having an understanding of what is going on inside your body often makes a tremendous difference in the handling as well as the outcome of the problem.

If any of these common problems occurred, would you know what to do about them?

- You are about to take a laxative from the medicine chest. As you reach for it, you stop yourself. You realize you are often constipated. Is it normal to be constipated? How often should you move your bowels? Will you be sick if you don't move your bowels for one, two, or even three days? What causes constipation, anyway? Is there anything you can do besides taking laxatives to help the problem? (See Chapter 3 for answers.)
- You are traveling far from home. Because Mexico is not on your itinerary, you are not worried about getting diarrhea. You plan to stay in only a few big cities in Europe. On the third morning of your trip you have diarrhea. What could it be from? Did you eat something that didn't agree with you? Should you go to the local drugstore and ask for some medication? Is your condition serious? Is there any way you can treat yourself? Should you fly home? (See Chapter 15 for answers.)
- After a tense and tiring day, you rush home, eat your dinner, and go right to bed. As you are falling asleep, you suddenly get a sharp pain in your chest, accompanied by a burning sensation. Are you having a heart attack? Should you go to an emergency room? Is there anything you can do at

home to relieve the symptoms? (See Chapter 8 for answers.)

- You've been under a lot of pressure lately. It does not come as a surprise to you when you develop what you think are typical ulcer symptoms. Instead of seeing your doctor, you decide to treat it at home. You drink a lot of milk and eat a lot of milk products. Initially, you seem to feel better, but a few days into your program, the symptoms worsen. You are puzzled and a little frightened. Why didn't the milk work? Could you have something more serious than an ulcer? Could you have cancer? (See Chapters 7 and 17 for answers.)
- Being concerned about your health and worried about getting older, you decide to start taking vitamin and mineral supplements. On the advice of some magazine articles you've read, you choose what you think is the proper dosage for yourself. Time passes and you begin to feel weak and tired, notice your appetite isn't what it used to be, have occasional headaches, and find your joints and bones ache. Could these symptoms have anything to do with the supplements you're taking? You wonder if you should stop taking them altogether, or perhaps eliminate some and double up on others. But then you reassure yourself that all vitamins and minerals are safe or they wouldn't be sold without a prescription. What should you do? (See Chapter 2 for answers.)

One of every six major illnesses involves the digestive tract. More than 5 million people each year are hospitalized for a digestive disorder. Untold millions suffer without seeking medical attention, as is indicated by the over half billion dollars spent annually on over-the-counter medications. Digestive problems account for the greatest amount of industrial absenteeism in the country—even more than the common cold.

Of all the systems in your body, your digestive system is most likely to give you trouble. And when it does go awry, your whole sense of well-being is disturbed. You may experience pain or discomfort, lose your appetite, feel nauseated, or vomit. You may become constipated, have diarrhea, or alternate between the two. Your skin may turn yellow, you may have bleeding or become anemic, you might become dehydrated or weak. You may have heartburn, feel dizzy, or even feel so sick you can't get out of bed in the morning. When your digestive system is in trouble, you're in trouble.

Almost anything can affect the way the digestive system functions, including your emotions, diet, sex life, job, genes, and the amount of alcohol and drugs you consume. In the past ten years, many new and exciting discoveries have been made in the area of digestive disorders. For instance, did you know:

- Why taking vitamin C can mask one of the major signs of bowel cancer?
- Of a food that can help restore normal bowel functioning?
- A few simple procedures you can do at home to help relieve heartburn?
- Why milk can actually be bad for an ulcer?
- Why the time antacids are taken can actually make a big difference in how effective they are?
- Why some ordinary over-the-counter medications can have a negative effect on the digestive system, causing ulcers, diarrhea, and constipation?
- How certain dietetic foods can cause bowel problems?
- Why jogging can make you constipated?
- Which foods to avoid, especially in the summer, to lessen the chances of getting food poisoning?
- That there are a few simple precautions travelers can take to prevent and treat diarrhea?
- Why having a cold or the flu can make you constipated?

Although this book contains advice about various

treatments, medications, exercises, and preventive measures, it is in no way intended to be a substitute for a doctor's advice. A personal physician is in the best position to diagnose and treat a particular problem. A doctor evaluates the symptoms, gives a complete examination, orders the appropriate tests, and then, like putting together the pieces of a jigsaw puzzle, makes an evaluation of the problem on the basis of all the findings. My intention in writing this book is to help readers become more active and knowledgeable participants in their own health care. Nothing is more valuable than good health. And being healthy is more than just being free of disease. It is a positive and joyful state of being, where every system in your body is working smoothly and at top capacity.

1

WHAT IS THE DIGESTIVE SYSTEM?

In the course of twenty-four hours, your heart beats over 100,000 times, pumping your blood more than 150 million miles. You breathe over 23,000 times, move about 750 different muscles, and speak about 4,800 words. You eat, digest, and absorb approximately 3½ pounds of food and drink two to three quarts of liquid. In order for these processes to take place, each organ in your body must be constantly nourished and fueled to create the energy needed to do its job. You could not live without a way to take in food and fluids and to break down their complex molecules into simpler substances that can be absorbed by your body. This is the vitally important work of the digestive system.

Assume you have just eaten a breakfast consisting of orange juice, two eggs, toast, and coffee. Although you are not really conscious of what is going on inside your body, an extremely important process is taking place. The digestive process begins as a voluntary one: you respond to the sensation of hunger by eating. Using your teeth, you begin to grind up the food and mix it with saliva. The grinding and moistening of the food makes it easier to swallow and at the same time brings the food into contact with the enzymes in saliva; those

What Is the Digestive System?

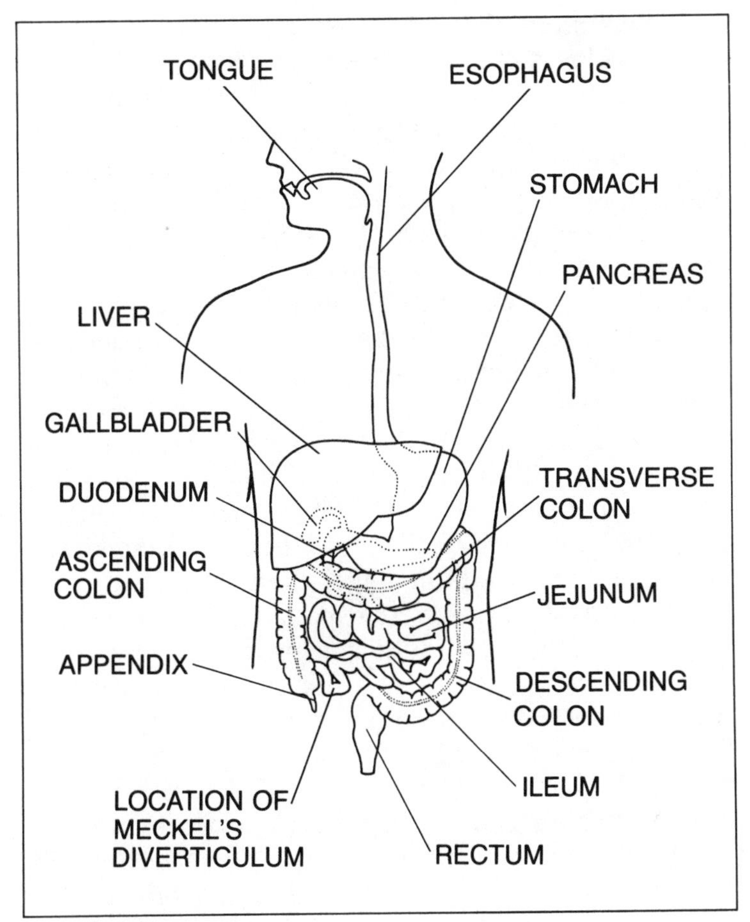

The digestive system.

enzymes begin digestion by starting to convert starches into the sugars you will need for the energy to continue digestion.

Still in control of what you are doing, you push the food to the back of your mouth. This triggers a contraction of muscles in the back of the throat and at the base of the tongue that pushes the food into your esophagus. From this point on, the digestive process becomes an automatic one. With the help of gravity and strong muscular contractions, the food is carried by your esophagus down through your neck and chest into your stomach. The esophagus, a ten-inch-long, pinkish-colored tube with the diameter of a half dollar, is extremely efficient. Its contractions are so strong you could drink a glass of water while standing on your head and the water would still reach your stomach. It is so elastic that people have been known to swallow extremely large objects without difficulty. Your esophagus ends with an opening at the top of your stomach called the lower esophageal sphincter, which opens to allow food into the stomach and, under normal conditions, immediately closes so food cannot return up the esophagus.

From the moment you began to eat, the stomach has been readying itself to make its own contribution. It is a J-shaped organ, pinkish in color, about ten inches long, and lies horizontally in the upper abdomen. Though it is only a little bigger than an adult fist, the stomach has the capacity to expand to hold as much as two quarts of liquid without causing any distress. It has several functions, acting as a storage area for your entire meal, as well as secreting digestive enzymes and hydrochloric acid to kill ingested bacteria, break up the tough, fibrous components of animal tissues, erode the cementing substances between cells of foodstuffs, and break down all the food you eat into a semisoluble substance called chyme. Although hydrochloric acid is strong, the stomach normally does not damage or digest itself because of its protective layer of gastric mucus.

Contracting about three times per minute, the stomach keeps mixing the chyme with the acid until slowly, after two to three hours, it begins to release the contents into the upper portion of the small intestine (the duodenum). Most of the digestive process occurs here, in the upper ten inches of the small intestine. As the chyme enters a little at a time, it stimulates the secretion of digestive juices by the pancreas, the liver, and the lining of the small intestine itself.

The small intestine is one of the most fascinating digestive organs. If it were uncoiled it would extend to over twenty feet in length and 1½ inches in diameter; if all the absorptive surfaces were flattened out, it would cover 5,400 square yards, or about the same expanse as a tennis court. The small intestine is responsible not only for digestion but also for the absorption of all essential nutrients into the bloodstream.

Secretions from the pancreas, which enter near the top of the duodenum via the pancreatic duct, are highly alkaline; therefore, they counteract and neutralize the hydrochloric acid coming from the stomach. They also contain powerful enzymes that contribute to the final breakdown of protein and carbohydrates and help to digest fats. Bile, manufactured by the liver, reaches the duodenum by a more circuitous route; it travels through ducts from the liver into the gallbladder and is stored there until it is needed. When dietary fat reaches the duodenum, the gallbladder contracts, releasing bile into this area. The bile surrounds the fat cells and emulsifies or dissolves them into a liquid that can be absorbed by the body.

While all of this is happening, the small intestine keeps churning and mixing the digestive enzymes with the chyme, assuring that absorption will take place. The contents of the small intestine keep decreasing as the soluble food is absorbed by the millions of fingerlike projections (villi and microvilli) lining its walls. By the time the food has reached the ileum (the end of the small intestine), all the nutrients the body needs have been absorbed from it. At this point, the connection between the small and large intestines (the ileocecal valve) opens, and the remaining residue is propelled into the colon (the large intestine).

All the remains of the food, other than water, are undigestible fragments of gristle, fiber, seeds, and small amounts of bile and salt. The colon's main job is to extract the water and salts from the material and prepare the residue for elimination. But even here, a lively population of intestinal bacteria attack the material in a final search for useful nutrients. These intestinal flora flourish in the colon because the warm, dark, moist atmosphere serves as an effective breeding ground for them. Here they multiply so rapidly that they make up 10 to 20 percent of solid stool. Some of these bacteria have no effect on the body at all, while others are extremely beneficial, producing more than half the vitamin K the body needs for blood clotting. The remaining 80 to 90 percent of the intestinal material becomes stool as water is drawn from it and it is pushed into the last five to six inches of the rectum. Here it is stored, lubricated, and made ready to be passed out of the body.

Not only is the digestive system busily at work twenty-four hours a day processing food and turning it into chemical substances that can be absorbed and used by the body, it is also standing guard to protect against any harmful toxins or organisms that may invade the body. Through both mechanical and biologic means, it provides a strong defense system.

Its first line of defense is actually the saliva which, when mixed with a toxin, will rapidly dilute it so the system can handle the toxin more easily. Any organism making its way down into the stomach will be met by a formidable enemy—hydrochloric acid. Many bacteria entering through the mouth in food are immediately destroyed by this powerful solution.

Should any harmful substances slip by into the small intestine, they again will be met head-on by a strong protective team. The small intestine has the incredible ability to recognize chemicals or bacteria harmful to the body. Often, it will try to rid the system of these enemies by a simple mechanical procedure, either by causing a vomit reaction or by coating the enemy in mucus and rapidly flushing it out of the system—diarrhea. If this doesn't work, it has an even more powerful means of dealing with the problem. Millions of cells lining the intestine are capable of producing antibodies to fight off invading bacteria. If all else fails, and bacteria or other pathogens overcome intestinal defenses to enter the bloodstream, the liver takes over as the next defense barrier. One of the many functions of the liver, which is a major component of the body's immune system, is to filter all blood coming from the intestine before it can flow to the rest of the body. During this process, special cells in the liver, called Kupffer cells, remove any remaining bacteria or antigens from the blood.

Because the entire digestive tract is constantly being assaulted by swallowed toxins and bacteria, the intestinal lining is often injured. In order to defend itself, it has the capability of regenerating the cells lining its walls; in just twenty-four to forty-eight hours, it can create a whole new lining to replace the injured one.

In any mechanism as complex as this one, there are bound to be problems. These problems, even the minor ones, can cause extremely unpleasant symptoms and interfere with the whole digestive process. These will be explained in the following chapters.

2
NUTRITION AND SELF-CARE

It may surprise you to discover that in many ways you're in control of how well you feel, how much you're able to enjoy your life, and how long you're going to live.

The value of prudent self-care has finally come to light. Researchers agree that what we eat, how much we exercise, and the way we handle stress all play a major role in our health and longevity. However, the battle rages on between the well-meaning health faddists, the out-and-out charlatans, and the ongoing scientific research being done in this field.

In this chapter, I am going to clear up many of the myths that prevail in the area of self-care, while at the same time providing you with the information you need to keep yourself healthy. You may discover that to maintain your health you will have to change a lifetime of bad habits. I know how difficult this can be, but the rewards are tremendous. Feeling well, having lots of energy, being sexually active, and living your life to the fullest should be enough motivation for you to change. Growing older is a privilege not afforded to everyone. Appreciate it and don't abuse it. If you follow a few simple rules, you can continue getting as much out of life at eighty as you did at eighteen.

How can I determine if the health information I get is valid?

Often, this is difficult. In the 1980s, everyone is a self-styled health expert. You may receive health-related information from various sources, some of which is valid, some of which is useless, and some of which is dangerous. Being able to discern fact from fiction in this area is extremely important. Be cautious. Your health is too important to you to take any risks with it. Be honest with yourself. If you are past fifty and/or have one or more chronic diseases, you are likely prey for the unscrupulous. Remember the adage: If it seems too good to be true, it usually is.

The most common area of health fraud is in the field of nutrition. A 1976 amendment to the Food, Drug and Cosmetic Act prohibits the FDA from regulating the potency of food supplements. This loophole allows almost anything to be marketed as a food supplement without having to be properly tested.

As an intelligent consumer, be aware of how a product is marketed. Be alert to promotional campaigns exploiting your fears and hopes, as well as ones promising instant cures. Any product that is supposed to be a "new scientific breakthrough" but is unknown to the medical community should immediately arouse your suspicion.

Most fraudulent products are advertised through the use of anecdotal statements from previous consumers. Don't be fooled by this. Also, watch out for anyone who tells you that whatever medical problem you have will be cured by proper nutrition. A good diet is certainly important, but thus far heart disease, cancer, and other serious diseases have not been cured by dietary nutritional means.

Should I discuss my diet with my physician?

Definitely. What you eat, how often you eat, and how much you eat are all valuable parts of your total medical history. Be honest with your doctor about your diet. Many people are embarrassed to admit they eat a lot of junk food, they may not be capable of preparing nutritious meals, or they feel their budget doesn't allow for a healthy diet. If you need some help in meal planning, your doctor may refer you to a nutritionist at a hospital in your area. Sometimes a few simple tricks in shopping or cooking can make all the difference in the world.

If your doctor knows you are missing a specific nutrient, vitamin, or mineral, it might provide him or her with a valuable clue to the cause of your symptoms. Don't forget that even your bowel habits depend a great deal on what you're eating.

Can taking vitamins and minerals without a doctor's prescription be dangerous?

Vitamin therapy is not without risk. Many people are fooled into thinking vitamins and minerals are harmless, regardless of the dosage, because they are sold in health-food stores. The average person doesn't think of them in terms of what they really are—chemicals altering the way your organ systems function. Vitamin supplements should be used only to correct specific deficiencies and only under medical supervision. Excessive and/or prolonged use of vitamins is of questionable value and is not without risk.

- **Vitamin D** is related to calcium metabolism, and most individuals require no more than 400 IU a day. Excess amounts of vitamin D can result in kidney stones and worsening of osteoporosis.
- **Niacin,** in doses as low as 750 mg a day, can cause toxicity after only three months. The toxic effects of niacin include abnormal liver function and jaundice. Also, rash, flushing, itching, and pigmentation may occur. Heartburn, nausea, vomiting, rapid pulse, drop in blood pressure, fainting, and elevated blood sugar are other manifestations of niacin toxicity.

- **Vitamin A** intoxication may occur in doses of 200,000 IU after two months. Doses as low as 50,000 IU a day may cause toxicity after more prolonged use. Vitamin A toxicity produces changes in the skin. Dryness, itching, rash, pigmentation, fissures, and desquamation of the skin all occur with excess vitamin A. In addition, hair loss, mental changes, bone and joint pain, muscle stiffness, fatigue, loss of appetite, and headaches occur. Endocrine abnormalities and enlargement of the liver or spleen may develop.
- **Vitamin E** is the subject of controversy and misunderstanding. Several studies have shown no statistical health benefit resulting from its use. Some side effects reported with vitamin E supplements in doses ranging from 400 IU to 2,400 IU include nausea, diarrhea, intestinal cramps, headache, flushing, weakness, and fatigue.
- **Vitamin C,** also known as ascorbic acid, is water soluble, and excess amounts are excreted in the urine. As little as ½ g can cause diarrhea. Four to 12 g a day can cause crystals of urate or cystine to precipitate from the urine and form painful kidney stones. Vitamin C may also cause painful irritation of the esophagus, especially when swallowed with inadequate liquid. This results in heartburn and, rarely, ulcers of the esophagus.
- **Thiamine** toxicity is rare, but allergic reactions may occur, including shock and cardiac abnormalities. Neurologic symptoms include seizures, trembling, and neuromuscular paralysis.
- **Vitamin B_6,** also known as pyridoxine, can cause convulsions when either too little or too much is taken. The specific dose-related symptoms are unknown.
- **Vitamin B_{15},** or pangamic acid, has been widely touted as a cure-all for all sorts of ailments, including cancer, diabetes, and hypertension. It has even been promoted to combat the effects of aging. In fact, pangamic acid is not a vitamin at all. There is

no evidence that it is of any value and it is not recommended under any circumstances.

Is it dangerous to take a multivitamin every day?
In most cases, a multivitamin is harmless. However, people with certain medical conditions may experience irritation of the digestive tract if they take daily multivitamins. Check with your doctor.

Is it true that in this country many older people are malnourished?
Inadequate nutrition in people over sixty-five is a problem, not just in this country, but all over the world. Statistics show that more than half of Americans over the age of sixty-five are not getting the minimum daily requirements of essential nutrients, vitamins, and minerals.

Many factors account for this. Because knowledge of the importance of nutrition is relatively new, many older people are not well informed about how to eat properly. As we get older, our life-styles change in a way that can seriously affect our diets. The loss of a spouse, suddenly having to live alone, a decrease in financial resources, ill-fitting dentures, chronic illnesses, and medications all play a role in our eating habits as well as in our ability to digest and absorb food. Aspirin may cause depletion of ascorbic acid (vitamin C) levels in older people. Excessive use of mineral oil laxatives limits the absorption of fat-soluble vitamins such as A, D, and K. Aluminum-containing antacids increase the excretion of calcium and phosphorus from the body and can worsen osteoporosis in older people. The use of diuretics may deplete the body of many important minerals, including potassium and magnesium.

Is there any way I can determine if I need vitamin and mineral supplements?
Yes. Keep a list for one week of what you are eating each day. When the week is up, study the list carefully and ask yourself these questions:

Nutrition and Self-Care

Is my diet balanced? Am I eating proteins, fats, and carbohydrates every day? Does my daily menu include fresh fruits and vegetables, grains, meats, chicken or fish, and dairy products? Are most of the foods I eat processed or just plain junk food? Is a TV dinner my most nourishing meal of the day? When you have answered these questions honestly, you will have a good idea of what your diet consists of and how it may be lacking important vitamins and minerals. Now ask yourself how you're feeling. If you're suffering from early vitamin and mineral deficiencies, the symptoms may be quite subtle, and you may think of them as a normal part of aging.

What symptoms should I look for?
If you are not feeling as well as you should be, lack energy, have a poor appetite, have unexplained weight loss, are depressed without knowing why, don't sleep well at night, or if you have trouble fighting off infections, you may be suffering from one or more deficiencies.

Is there a way to scientifically establish a deficiency?
Yes. There are specific blood and urine tests that can determine if you have too much or too little of any of the nutrients, vitamins, or minerals.

If my doctor finds a deficiency, how is it treated?
That depends on the underlying cause. To determine the cause, your physician will review your diet, check the medications you are taking, and investigate to see if there are any deficiencies in absorption and digestion.

What is malabsorption?
This is a broad term used to explain any condition associated with a decrease in the ability of the body to absorb nutrients. Maldigestion is included under this heading. This usually results from a defect in the release of any of the digestive enzymes, particularly the

pancreatic enzymes, which are needed to digest protein, carbohydrates, and fats. Bile made in the liver and stored in the gallbladder is also needed for the digestion of fats. If bile production is interfered with because of damage to the liver, maldigestion will result.

Malassimilation is another cause of malabsorption. This can occur when a surgical procedure interferes with the normal flow of nutrients through the small intestine and their adequate mixing with the digestive enzymes; this may happen after surgery for ulcer disease. Fortunately, the condition is rarely severe enough to cause serious problems.

Why is calcium so important as we get older?

The major mineral component of bone is calcium, and it decreases with age. Excessive loss of bone calcium leads to osteoporosis, a condition responsible for many of the fractures and spinal disorders of older people. In men above the age of fifty, bone minerals decline by only ½ percent a year, and osteoporosis rarely occurs before the age of eighty. In women, on the other hand, 1 percent of bone calcium is lost each year beginning at age thirty-five; after menopause, with the decline in female hormone levels, this rate increases to 5 percent. By the age of sixty, more than 25 percent of women have some degree of osteoporosis. Reduction in physical activity plays a major role in bone calcium loss in both men and women, and is a major factor in the development of osteoporosis in older people.

Is it only calcium that prevents osteoporosis?

No. Many nutrients interact to ensure healthy bones, including magnesium, zinc, phosphorus, fluoride, protein, vitamin A, and vitamin C. With these nutrients, specific hormones, including estrogens, calcitonin, and parathyroid hormone, play their role.

Although osteoporosis is irreversible, it is also a progressive disorder. For this reason, it is essential that you not treat yourself without first consulting your physician. He or she is best able to integrate the many factors involved in osteoporosis as they pertain to your particular age and state of health. Your physician can help you plan a sound dietary regimen and exercise routine, and aid you in choosing the proper supplements to offset the progression of this disease.

How can I get adequate calcium from my diet if I cannot tolerate milk?

Lactose (milk sugar) can only be absorbed in the presence of the intestinal enzyme, lactase. This enzyme is always present at birth, but may begin to drop off after the weaning period, producing a lactase deficiency. This deficiency is particularly common among Greeks, Jews, Orientals, Blacks, and Native Americans.

If the milk sugar is not properly absorbed by the small intestine, it passes down into the large intestine and causes excessive gas, bloating, and diarrhea.

Because so many people do have a problem digesting milk, it is important, especially for women over forty, to get their calcium intake from other sources. Sardines and canned salmon, particularly with the bones left in, broccoli, kale, oysters, and dandelion, turnip, and collard greens are also good sources of calcium. Interestingly enough, the bacteria in some types of yogurt produce their own lactase; this enables people who cannot digest milk products to eat yogurt without suffering digestive problems.

Calcium supplements may be taken in the form of calcium tablets available in most drug stores, but they should not be taken without consulting a physician.

What is a "good" diet?

There are some basics to a "good" diet that I will outline here, but for a diet plan to be beneficial, it must be individually designed to take into consideration any specific needs or restrictions you may have. Foods should be chosen from the four main food groups: dairy products; meats, fish, and poultry; fruits and vegetables; and grains (which cover a far broader area than just breakfast cereal). Ideally, foods from each of these groups should be included in your daily diet. Also, your diet should strike a balance between proteins, carbohydrates, and fats.

Proteins are the building blocks of all the tissues in the body. They also make up the enzymes that enable the life-sustaining biochemical reactions to take place. Lack of protein creates serious problems, including lowered resistance to infection, an impaired nervous system, diminished utilization of intellectual capacities, and rapid aging. Serious deprivation may shorten one's life span. The best sources of protein are lean meats, poultry, seafood, and fish.

Carbohydrates are energy foods. They can either be eaten as simple sugars, or as starches that are broken down into sugars. Bananas, potatoes, corn, rice, pasta, kidney beans, honey, and table sugar are all rich sources of carbohydrates.

Fats, both saturated and unsaturated, are necessary to provide body heat, help us utilize carbohydrates, and provide the fatty covering for our nerve cells. The body stores fat to be used as an emergency source of food when we are not eating. Fats should be limited in our diets because, ounce for ounce, fat contains more calories than proteins or carbohydrates. But they should not be eliminated altogether.

Much has been said and written about the importance of limiting fats, especially saturated fats, in the diet. Experts have long felt that saturated fats were a contributing cause of heart disease. Although this is still a subject of debate, it is probably advisable to limit the amount of saturated fats in the diet.

What do vitamins and minerals really do?

Each vitamin and mineral works to ensure the smooth functioning of our organs, as well as contributing to the biochemical reactions going on in our bodies. The following list explains the functions of many vitamins and minerals.

- **Vitamin A** is essential for healthy skin, eyes, teeth, gums, hair, and for adequate vision in dim light.
- **Vitamin B_1** (thiamine) is needed for the metabolism of carbohydrates and energy-producing reactions in the body.
- **Vitamin B_2** (riboflavin) is necessary for healthy skin. It helps prevent sensitivity of the eyes to light, helps the body cells use oxygen, and is essential for the structure and maintenance of healthy body tissues.
- **Vitamin B_6** (pyridoxine) permits the formation of normal red blood cells and helps in the metabolism of the amino acids that make up body proteins.
- **Vitamin B_{12}** helps prevent certain forms of anemia. It contributes to the health of the nervous system and to the formation of normal red blood cells.
- **Folic acid** helps in the formation of red blood cells and in the prevention of certain forms of anemia.
- **Pantothenic acid** is essential for the metabolism of carbohydrates and fats.
- **Niacin** is essential for converting food to energy. It is necessary for the formation of certain hormones and for a healthy nervous system.
- **Biotin** is needed to form fat-like substances in the body.
- **Vitamin C** (ascorbic acid) is essential for normal teeth, bones, and blood vessels, and for the formation of collagen, a protein that helps support body structures.
- **Vitamin D** is necessary for teeth and bones. It helps in the utilization of calcium and phosphorus and prevents rickets.
- **Vitamin E** is needed for normal muscle, red blood cells, and other chemical constituents of the body tissues.
- **Vitamin K** is essential for normal blood clotting.

Below I have listed foods providing the richest sources of these vitamins:

- **Vitamin A:** Milk, butter, fortified margarine, eggs, liver, green leafy vegetables, yellow vegetables
- **Vitamin B_1** (thiamine): Enriched bread, fish, lean meat, liver, pork, poultry, dried yeast, milk
- **Vitamin B_2** (riboflavin): Eggs, enriched bread, leafy green vegetables, lean meats, liver, milk
- **Vitamin B_6** (pyridoxine): Vegetables, meat, whole grain cereals
- **Vitamin B_{12}:** Liver, kidneys, milk, salt-water fish, oysters, lean meat, foods of animal origin in general
- **Folic acid:** Leafy green vegetables, liver, eggs
- **Pantothenic acid:** Eggs, nuts, liver, kidneys, green leafy vegetables
- **Niacin:** Lean meats, liver, dried yeast, enriched bread, eggs
- **Biotin:** Liver, kidneys, eggs, most fresh vegetables
- **Vitamin C:** Citrus fruits, berries, tomatoes, cantaloupe, potatoes, green leafy vegetables
- **Vitamin D:** Vitamin D fortified milk, cod liver oil, salmon, tuna, egg yolk
- **Vitamin E:** Vegetable oils (cottonseed, olive, margarine), whole grain cereals, lettuce, walnuts, almonds
- **Vitamin K:** Leafy vegetables

The following is a list of minerals essential to your health.

- **Calcium** is needed to keep bones strong. It is involved in blood clotting and is essential for maintaining a normal heart rhythm.
- **Chromium** helps break down fats and cholesterol in blood. It is also essential for the body's utilization of carbohydrates.
- **Copper** helps in the manufacture of hemoglobin, the part of blood carrying iron.
- **Iron** is necessary for the production of hemoglobin. It helps the body resist disease and stress.

- **Magnesium** is essential to the proper functioning of nerves and muscles.
- **Manganese** helps maintain the production of sex hormones. It is important for the proper functioning of the brain and nerves.
- **Phosphorus** is needed for cell repair and energy production. It is also essential for healthy kidney functioning and the smooth workings of the nervous system.
- **Potassium** is found mostly inside of body cells and aids in the proper maintenance of cellular function. It works with sodium to regulate the salt and water balance.
- **Selenium** is necessary for normal body growth, metabolism, and fertility.
- **Zinc** is necessary for healing and for the formation of new cells. It aids the digestive enzymes and is important in maintaining the prostate gland.

Here is a list of foods richest in minerals. Use it to help make your dietary choices in the same way you use the list of the high-vitamin foods. You will notice many of the foods are rich in both vitamins and minerals, which should make your choices easier.

- **Calcium:** Milk and milk products, soybean curd
- **Chromium:** Meats, whole grains, cheese, brewer's yeast
- **Copper:** Liver, legumes, eggs
- **Iron:** Liver, eggs
- **Magnesium:** Nuts, beans and peas, soy products
- **Manganese:** Tea, wines, unrefined cereals, herbs, spices
- **Phosphorus:** Meat, fish, poultry, eggs, nuts
- **Potassium:** Green leafy vegetables, oranges, potato skins, bananas
- **Selenium:** Seafoods, meats, cereals
- **Zinc:** Beans, nuts, liver, red meats, oysters, wheat germ

Can medication affect how the body utilizes vitamins and minerals?
Every drug, whether it is prescribed by a physician or purchased over the counter, affects our bodies in some way. Some directly interfere with vitamin and mineral absorption, and others create a need for extra amounts of specific vitamins and minerals. Some medications push these substances out of the body so quickly they do not have a chance to be utilized at all.

Which common medications affect the utilization of vitamins and minerals?
Aspirin and tetracycline (an antibiotic) actually block vitamin C from entering the bloodstream. Phenobarbital, aspirin, and triamterene (a diuretic) affect the way the body utilizes folic acid. Hydralazine (an antihypertensive) and L-dopa (a drug used to treat parkinsonism) often cause a vitamin B_6 deficiency. Mineral oil, a commonly used laxative, depletes the body of carotene, vitamin A, vitamin D, and vitamin K. Antacids that contain aluminum inhibit the absorption of phosphorus and increase the excretion of calcium. All diuretics work by pushing fluids out of the body quickly. At the same time, they also drain the body of many needed minerals, such as potassium.

Does radiation therapy cause any specific nutritional problems?
Yes. If radiation is given to the area below the waist, radiation proctitis, radiation colitis, or radiation enteritis can develop, either during the treatments or after the treatments are completed, sometimes even years later. Usually, these problems are self-limited, but sometimes may require hospitalization. If radiation is given to areas above the waist, there is the possibility of damage to the esophagus. When this happens, there may be pain with or without swallowing and difficulty eating solid foods. However, this problem can also be treated. Usually, the esophagus is stretched by special

instruments; in some cases, surgical correction is necessary.

Which diseases of the digestive system can cause nutritional deficiencies?
Because the small intestine is responsible for absorbing nutrients taken in, any diseases in this area of the body may cause serious nutritional deficiencies. These diseases include, for example, Crohn's disease (see p. 108), radiation enteritis, sprue, lymphoma, giardiasis (see p. 142), and pernicious anemia.

Are there any general statements you can make about vitamin and mineral needs in people over fifty?
Yes, although I always hesitate to make general statements about nutrition because nutritional needs vary greatly from one individual to another.

Studies indicate that women over fifty are likely to be low in the following nutrients: calcium, iron, magnesium, vitamin B_6, folate, and zinc. They also may be low in vitamin A, riboflavin, and dietary fiber. One positive note for women in this age group is that, after menopause, when monthly bleeding ceases, there is less need for iron.

Whatever your age or sex, it's a good idea to consult with your physician about your specific nutritional status and needs.

Should I drink eight glasses of water a day, or is that an old wives' tale?
It may be an old wives' tale, but in this case, those old wives knew what they were talking about. You can live weeks without food, but if you go more than 24 to 36 hours without water, your system can become seriously dehydrated. Water is a necessary solvent for all the chemicals in your body, and is a major component of every living cell.

But as vitally important as water is, the body has no way to store it. Each day, the fluids you take in are used up and the excess excreted, and you are left needing just as much the next day. Depending on the climate and your rate of activity, two and a half to five quarts of water are lost daily through urination, perspiration, and respiration. Vomiting and diarrhea rapidly force even more fluids from your body. This can result in dehydration, a life-threatening emergency, especially in children and in elderly persons. If you take in more than you need, you will just urinate more frequently. The best advice is to maintain good hydration by drinking adequate amounts of water daily.

What role does exercise play in maintaining good health?

Exercise is important in maintaining good health. A sensible exercise program, designed specifically for the needs of the individual, can markedly slow down any loss of physical strength and at the same time provide a sense of fitness and well-being.

Every exercise program should begin with a complete physical examination so the physician can recommend the type and amount of exercise most beneficial for the individual without creating any risk factors. Crash programs or exercise fads can be extremely dangerous, particularly to people over fifty. A good program will call upon the physical reserves of the person gradually, building up over an extended period of time. Sudden overexertion can be deadly. However, it is just as dangerous not to exercise at all. Inactivity leads to obesity, flabby muscles, diminished breathing capacity, loss of appetite, loss of bone minerals (especially in older people), constipation, and impaired cardiovascular functioning.

If you are not exercising now, it's time to begin. Though no exercise program can help you recapture the stamina of your youth, you can certainly minimize the hazards of inactivity and increase your sense of well-being. (One good resource book for exercises is *Fitness for Life: Exercises for People Over 50,* Theodore Berland [AARP Books, 1986].)

Nutrition and Self-Care

Are there any dietary secrets to preventing cancer?
Many food supplements, vitamin and mineral combinations, and exotic potions are purported to be able to prevent cancer, but so far no such claim has been scientifically proven. In studies over the years, it has been shown that people on high-fiber diets low in animal fat seem to be at reduced risk of developing colon and breast cancers.

Is smoking as dangerous as they say?
Absolutely! All research indicates that smoking causes multiple health problems. The only positive thing to say about it is that it brings in billions of dollars a year for the tobacco industry.

Is drinking bad for your health?
The answer to this depends a lot on the kinds of health problems you have. I would suggest you discuss this question with your doctor. If your physician feels there is no medical reason for you not to drink, then a glass of wine with dinner may actually be good for you. Studies show that alcohol, in moderation, reduces stress and may be beneficial.

Is there anything else?
There is no doubt that diet and exercise play a significant role in both health and longevity. But they are not the whole answer. Much depends on our state of mind. How well we cope with stress, the loss of a loved one, financial reversals, illness, and loneliness all influence the way our bodies function. The strong connection between the mind and the body is now a well-accepted scientific fact.

Current research is exploring how our emotions affect the various chemical reactions in our body. When we have a positive attitude, we can keep our sense of humor even when things are not going well; our brains trigger chemical reactions that are beneficial and help our bodies function to their maximum capacity. Conversely, depression, anger, stress, and anxiety are detrimental. If you find there are areas of your life you have trouble coping with, I would recommend seeking professional help. There are many agencies offering both individual and group therapy at reduced cost. You can contact your local mental health association for the agency nearest to your home.

3
CONSTIPATION

Although constipation is the most common of all digestive complaints, it has been so clouded in misinformation that the average person has little understanding of what it really is or how best to deal with it. Over the years, constipation has been blamed for everything from irritability and headaches to total autointoxication.

The ancient Egyptians believed that constipation was the underlying cause of all disease. At that time, doctors would routinely give harsh laxatives with other medicines. And, constipated or not, the Egyptians spent three days each month undergoing heavy purges as a form of preventive medicine.

In the twelfth century, purge houses were established in England for wealthier citizens. They would visit them once a week to have their bowels cleansed by strong laxatives and enemas. It became such a fashionable and accepted activity that people would plan their social calendars around their purge days.

We no longer have purge houses, but many myths prevail similar to those that encouraged purging. You only have to open a magazine or turn on a TV to see how much of this kind of thinking continues. There is an endless array of advertisements for laxative products, all of which attempt to "cure" not only constipation but also a variety of symptoms that supposedly accompany it. Constipation is a real problem, but, under normal circumstances, not a serious condition. With a little knowledge you can learn how to handle it effectively and possibly even avoid it altogether.

The question-and-answer section that follows will help you learn the facts about constipation and will uncover some of the fictions surrounding this digestive disorder.

What is constipation?

Constipation is often incorrectly defined. Missing one or two bowel movements, or having unusually hard stools one day, is not true constipation. Real constipation is not an occasional event. A person who is constipated will go for days without having a bowel movement, and then will have very hard stools.

It is important to keep in mind that each person has an individual bowel pattern. Not everyone has a bowel movement every day; some people normally have bowel movements only two or three times a week. These people are not constipated. It is only when a person's bowel pattern changes from its norm for a period of time that the condition can be called constipation.

Is constipation a disease?

Constipation is a symptom, not a disease. It is the result of another problem.

Can constipation make a person sick?

Because of the way our bodies are made, it is impossible (under normal conditions) for any of the fecal material that isn't expelled from the colon to go back up the digestive tract and do any damage. In an otherwise healthy person, constipation in and of itself is not dangerous. But if you are frequently constipated or notice a sudden change in your bowel habits, there may be another condition causing the problem. If this occurs, you should discuss it with your doctor.

What produces the urge to have a bowel movement—and what happens if the urge is ignored?
When the stool moves down into the lower part of the colon, it creates a sensation of fullness in the rectum. This feeling is interpreted as the urge to move the bowels. If, on occasion, you don't have the time or the opportunity to go to the bathroom, no harm will come of it. But if you frequently delay defecation, problems can result. The longer the stool remains in the lower colon, or rectum, the harder and drier it becomes—and dry stool is one of the main causes of constipation. Furthermore, if you habitually ignore the urge, the defecation reflex is depressed until eventually you will be totally unaware of the need to defecate. No matter how busy you are, you should routinely set aside a time to go to the bathroom. This trains the bowel and gives a rhythm to the digestive system.

How can a person tell if a bowel movement is completed?
All stool that is passed during a bowel movement comes from the rectum. Once the rectum is emptied, there is a sensation of complete evacuation. If you still feel the urge to defecate, or continue to have the sensation of fullness and there does not seem to be any more stool to pass, it may indicate a hemorrhoidal problem, or it may mean that the rectum is inflamed—a condition known as proctitis. Most of the time, once evacuation begins, the entire rectum will empty.

If I have trouble moving my bowels, should I try straining?
Absolutely not! This can be dangerous. Straining at stool may strongly affect blood pressure and the heartbeat. In a person who is older and debilitated, or in someone who is predisposed to heart attack or stroke, this maneuver could actually bring on such a crisis.

Straining will also increase hemorrhoidal problems and can even cause the entire rectum to bulge out through the anal opening. It will also aggravate all types of hernias and force them to protrude. If you cannot move your bowels easily on any occasion, leave the bathroom until the urge to defecate returns.

What are the most common causes of constipation?
Improper diet and inadequate fluid intake definitely top the list. Most people don't realize how important dietary factors are to the proper functioning of the digestive system. In fact, under normal conditions, if there is an adequate amount of fiber and fluid in the diet, constipation can be eliminated. Fiber, also called roughage or bulk, helps promote the synchronous, wavelike contractions that keep the food debris flowing down the intestine. In addition, it expands the colon, facilitating the passage of fecal waste. As nondigestible fiber substances move through the intestine, they absorb over seven times their weight in water, resulting in softer, bulkier stools.

Many studies have been done on the effects of fiber in the diet, and they have all come to similar conclusions. A high-fiber diet helps to speed the wastes through the digestive system. Rural Africans, whose diets are high in fiber, eliminate their wastes in one-third the time that we do. Because of this, they experience almost no constipation and suffer from far fewer digestive problems in general.

History also provides evidence for the importance of fiber. In 1870, the roller milling machine was invented to refine wheat and remove most of its fiber content. Since then, medical literature abounds with reports of diverticulosis (the development of saclike bulges in the intestinal wall) and assorted problems in bowel function.

Which high-fiber foods are especially recommended?
The foods with the highest fiber content are bran and

whole-wheat products; fruits, including skin and pulp; green, leafy vegetables such as spinach, lettuce, and celery; root vegetables such as carrots, turnips, and potatoes; and the high-residue vegetables such as cabbage. Refined grains and processed foods should be replaced with these high-fiber foods.

Are there any other causes of constipation besides inadequate fiber and fluids?

Many factors play a role in the smooth passage of stool. A system that does not have enough exercise slows down. If you are under emotional stress, the tension can be carried over to the bowel and temporarily put it out of commission. If you have a cold, a flu, or any feverish illness, you will become temporarily dehydrated, and that may lead to constipation. Note that the key word here is "temporarily." As soon as health returns to normal, so will bowel habits. However, there is a more chronic and recurring kind of constipation that is usually caused by a digestive disorder. This is called the "irritable bowel syndrome" and it commonly causes the kind of constipation that recurs over many years. (For more information, see Chapter 11.)

Are there any substances that can cause constipation when ingested?

Yes. Many prescription drugs can cause constipation. Strong painkillers containing codeine slow the muscular contractions of the large intestine. These contractions are essential for rapid expulsion of the fecal material. It is well known that iron supplements are constipating. Antidepressants (Elavil®, Tofranil®, and Triavil®, to mention just a few) can also be constipating, and people receiving any of these drugs should be aware of this effect because these medications usually must be taken for prolonged periods of time. Mild tranquilizers (Valium® and Librium® are two common examples) may be constipating to some people, but in general are well tolerated. Alcohol, coffee, and tea, when taken in large quantity, can contribute to constipation by a diuretic effect of depleting body fluids.

Are older people more likely to have a problem with constipation?

Yes, but not for the reasons you might think. Laxative advertisers place the blame on the slowing of the digestive system that can occur with age. But such slowing is not inevitable—it is related to the decreased level of exercise that often accompanies aging—and is not the reason many older people become constipated. The real reason is that many older people have been laxative and enema abusers all their lives and, as they age, the practice begins to catch up with them. After many years of being stimulated unnaturally, the bowel loses its ability to contract and propel its contents as it normally should.

Constipation does not have to be a condition of aging. Many people in their eighties and nineties have smoothly functioning digestive systems, and rarely, if ever, suffer from constipation.

When should constipation be discussed with a doctor?

It is always wise to let your doctor know about any problems you have, including constipation. Though most constipation results from poor dietary habits, in a particular case, there may be a different underlying cause, such as another digestive problem or a problem in another part of the body.

By and large, if you normally defecate only two or three times a week, there is no cause for worry. This is how your digestive system works. But if your normal bowel pattern changes, it is extremely important to tell your doctor. Any change in bowel habit can signal a serious disease. Cancer of the bowel, for example, is often first discovered when a person notices a change in bowel pattern.

Should a laxative ever be used?

There are occasions when a laxative is the treatment of choice, but these occasions are rare. If you have experienced a temporary change in life-style, such as a vacation, an emotional crisis, or an extended period of bed rest after an illness, then a laxative may be helpful. If you do take a laxative, it should be an extremely mild one, such as Milk of Magnesia™. Once you return to your normal dietary and exercise pattern, your bowel problem should resolve itself.

People are not aware of just how harmful some laxatives can be and tend to take them without much thought. But laxative use can become laxative abuse all too easily. When it does, it makes it harder to treat the underlying problem.

If you suffer from chronic constipation, take a good look at your living habits. Does your diet contain enough fiber? Are you drinking enough liquids? Are you getting enough exercise? Are you under too much stress? If you can't locate the cause, or easily remedy it, I would advise you to discuss the problem with your physician.

What are the differences between laxatives and how do these products work?

Basically, there are four different kinds of laxatives. Irritant laxatives, the most popular ones, act as direct irritants to the bowel, causing it artificially to increase its motility. Bisacodyl (Dulcolax®), cascara sagrada, senna, and castor oil act in this way. Despite their popularity, these products have major drawbacks. Often, they will induce cramps and discomfort; even more important, if there is another problem present, they can do serious damage. In the presence of appendicitis, for example, an irritant laxative can prove fatal. This type of laxative should be totally avoided unless prescribed for a specific reason by your doctor.

The other three kinds of laxatives, osmotic laxatives, stool softeners, and bulk laxatives, are safer if used correctly. Osmotic laxatives work by holding on to water in the digestive tract, making the stool soft and moist. Milk of Magnesia™ is one of the better-known osmotic laxatives and is the one most commonly prescribed by physicians. However, even the osmotic laxatives can be dangerous if you have a preexisting water retention or kidney problem.

Stool softeners (Colace®, for example) work in a similar way, but instead of holding on to the water in the bowel, they actually force the water to penetrate the stool. (One word of caution here: Mineral oil, an emollient-type of stool softener, is an old home remedy. It is usually thought of as gentle, but it may actually be harmful. With prolonged use, it may deplete the body of the fat-soluble vitamins A, D, E, and K.)

Bulk laxatives (Metamucil® and Effersyllium®, for example) are made up of vegetable fiber and act in much the same way as fiber in the diet. In general, a high-fiber diet is recommended as the safest means of avoiding problems with constipation, especially among older people and people who are immobile or confined to bed. Fiber intake may be in the form of natural bran or commercially available bulk laxatives such as Metamucil® or Effersyllium®. They are the slowest acting of all the laxatives but, in the long run, the safest.

Are enemas recommended?

Unless a person is undergoing a diagnostic procedure, such as a barium enema or a colonoscopy, enemas are generally not recommended. Enemas have many hidden dangers. If the nozzle is not inserted carefully, it can injure the lining of the rectum. Soap, which is probably one of the most common home treatment ingredients, is an irritant to the bowel. If too much liquid is forced into the bowel all at once, it can injure the lining of the intestine, as well. Most people do not realize that the purpose of an enema is not to flush out the entire colon but to gently distend the rectum so that the normal reflexes can be set into motion. An enema may be the only way to relieve a fecal impaction or prepare for a diagnostic procedure. Certain precautions should be kept in mind: Only plain tap water or a mild commercial enema such as phosphosoda (Fleet®, for example) should be used, and the enema tip should be well lubricated and inserted gently to avoid traumatic injury to the rectum. Before using an enema, consult with your physician to be sure the indications are clear and the technique is understood.

What effect does exercise have on constipation?

That depends on the type of exercise. Any exercise that you do, a jog in the park, a game of tennis, or just some sit-ups in the morning, may cause you to work up a sweat. Sweating draws fluids from your body and may dehydrate you to some extent. This can have a detrimental effect on the consistency of your stool. When you exercise, it's important to keep this in mind and to replenish your fluid loss.

Some exercises will specifically help relieve constipation. Any exercise that strengthens the abdominal muscles makes it easier to pass stool. During defecation, these muscles are voluntarily contracted as a means of increasing intraabdominal pressure. This pressure helps push the stool out of the body. The stronger the muscles are, the more efficiently they work.

4
HEMORRHOIDS

A friend may corner you at a cocktail party to discuss her allergies or may freely delve into the mysteries of her psyche over dinner, but when it comes to a hemorrhoidal problem, she will likely fall silent. Most people are reluctant even to admit they have hemorrhoids. Yet, they are probably one of the most common disorders, affecting more than 75 percent of the population and causing an untold amount of pain and suffering.

If you are one of the 150 million people who suffer from this problem, don't despair. As you begin to understand them better, you will learn to handle hemorrhoids easily and effectively, eliminating a lot of unnecessary misery.

What causes hemorrhoids?

The rectum is endowed with a rich blood supply confined within a complex network of veins called a venous plexus. The term "hemorrhoid" is used to define any saclike dilatation or protrusion of these rectal veins. Certain factors contribute to the development of hemorrhoids, a prime one being posture. Because humans walk on two feet rather than four, continuous pressure is placed on the delicate venous structure in the rectum. The structure of the rectal veins in this area is somewhat different than that of veins in other areas of the body. They do not have any valves—a characteristic that increases the amount of pressure they must withstand.

Many environmental factors also play a role in the development of hemorrhoids. Obesity, the lifting of heavy objects, and continuous standing or sitting all add pressure to the rectal venous structure.

The major factor contributing to a hemorrhoid problem is constipation. Nothing puts more pressure on the rectal area than the continual passage of hard, irritating stools accompanied by prolonged straining. People who tend to be constipated also tend to be laxative abusers. The use of laxatives over a prolonged period of time weakens the supporting anal and rectal muscles that help keep the rectal veins intact. Although not everyone who is constipated will develop hemorrhoids, it's safe to say that constipation added to any of the other factors is enough to precipitate a hemorrhoid problem.

Are all hemorrhoids the same?

No, there are actually two different kinds of hemorrhoids and they produce two different kinds of symptoms. Because both kinds are usually present together, there is often confusion as to which hemorrhoid causes which symptom.

External hemorrhoids, which are sometimes referred to as "piles," form outside the rectum. These hemorrhoidal veins are covered with skin. If they swell or if a clot forms within them, their skin cover is stretched, causing the area to become painful. When the swelling goes down or the clot is absorbed, the empty skin tag remains.

Internal hemorrhoids are somewhat different. They are hemorrhoidal veins that are inside the rectum, underneath the rectal mucosa. Because they are not lined with skin, and there are no cutaneous nerve endings in that area, they usually do not cause any pain. But these

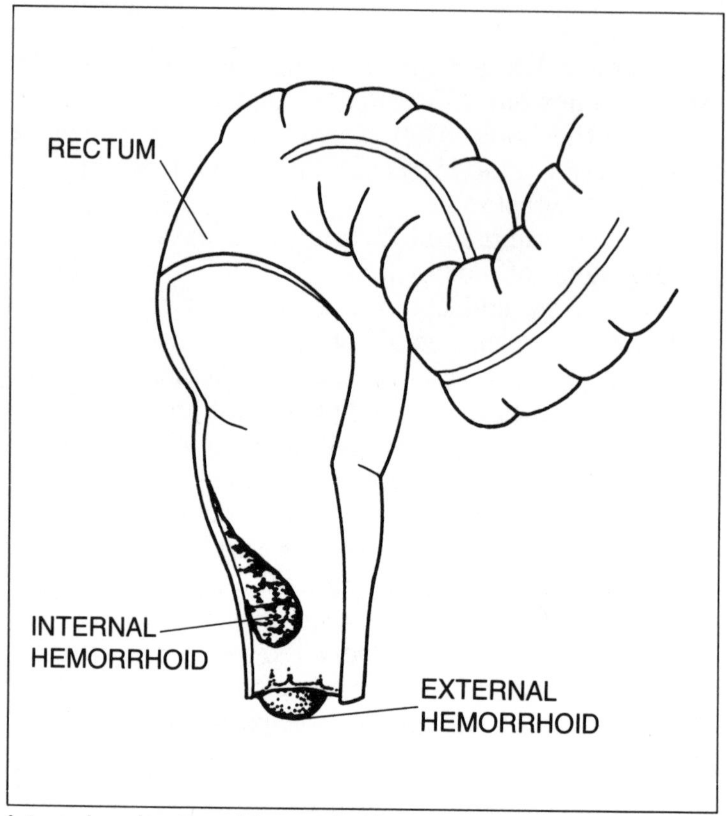

Internal and external hemorrhoids.

veins can also become swollen and even protrude outside of the rectum. They tend to bleed and are often the source of bright red blood on toilet tissue or in the bowl.

Can anyone get hemorrhoids?

Hemorrhoids are quite common in the adult population and symptoms may occur at any time. They occur in women during pregnancy, but are usually transient. Hemorrhoids also commonly occur in people who must stand for prolonged periods of time or who must lift heavy objects. They seem to be more prevalent in people with constipation or other difficulties with bowel movements. Many people over the age of fifty complain of irregular bowel habits, often use laxatives and enemas, and strain excessively during defecation. These habits only contribute to the hemorrhoid problem and exacerbate the symptoms.

How can I tell if I have hemorrhoids?

The four most common symptoms of hemorrhoids are similar to the symptoms produced by other disorders of the digestive tract, including cancer. If you think you have hemorrhoids, bring the problem to the attention of your doctor. Don't try to be your own physician and diagnose them yourself. If your diagnosis is incorrect, you may be ignoring a more serious problem and delaying necessary treatment.

The early stages of hemorrhoids are usually symptom-free. Most people do not even know they have hemorrhoids until either their doctor tells them or the problem has gone on for some time. As the problem progresses, depending on the type or types of hemorrhoids involved, the symptoms that develop may include pain in the rectal area, itching, the discovery of a lump around the anus, or the sighting of blood in the bowl or on toilet tissue. There also may be a sensation of fullness in the rectum or the feeling that you have not completed evacuation after a bowel movement.

What causes a sudden, painful attack?

When a blood clot forms within the walls of an external hemorrhoidal vein, it causes the vein to enlarge and swell. Because all external hemorrhoids are covered with skin, the skin must stretch to accommodate the swelling. This stretching is what causes the pain. Sometimes, irritation and tenderness will occur from the passage of a hard stool or from straining. But this discomfort is temporary and will usually go away by itself.

What should I do when I get a painful flare-up?

Once a doctor has confirmed your hemorrhoid problem, there are many ways to obtain quick and satisfactory relief when symptoms occur. The first order of business is to try to dissolve the clot and, at the same time, soothe the irritated tissues.

During a flare-up, maintain enough fiber in your diet to facilitate a rapid bowel movement. Excessive fiber adds bulk to the stool, whereas insufficient fiber will do just the opposite. Either way, defecation is prolonged, there is straining at stool, and hemorrhoids are irritated. The right amount of fiber in the diet can be determined only by trial and error. If your stools are too bulky, cut back on the fiber; if your stools are too small and segmented, add more fiber to the diet.

Drink plenty of fluids to keep stools soft. Stool softeners containing docusate (Colace®, for example) are helpful. Combination laxatives containing additional ingredients such as senna act as stimulants and are best avoided. Emollient suppositories (Anusol®, Calmol-4™, and Preparation H®, for example) may be used before and after bowel movements to lubricate the anal canal and relieve irritation.

Hygiene is also important. After you have a bowel movement, cleanse the area with gauze pads soaked in warm water and a mild soap solution or astringent. Then gently pat yourself dry. Never rub or scrub.

A warm sitz bath is an old home remedy that really works. Take the largest bath towel you can find and fold it until it makes a comfortable cushion for you. Place it in the bottom of a tub filled with about six inches of warm water. Sit there for 20 to 30 minutes, three to four times a day. This will not only soothe the inflamed tissues, but will also relax the spasms of the anal and rectal muscles that accompany the inflammation.

If you spend the better part of the day sitting, buy yourself a doughnut cushion to sit on. It is a round, inflatable, plastic cushion with a hole in the middle. It looks something like a child's swimming tube. When you sit on it, the inflated part bears the brunt of your body weight and relieves the pressure on the affected area. Also, try to lie down as much as you can (if you work, see if there's a couch in the office you can requisition during lunchtime). Standing also puts extra pressure on the sensitive hemorrhoidal veins.

If your internal hemorrhoids are swollen, you may not realize when defecation is complete. As a result you may tend to stay on the toilet longer than you should, aggravating the problem. Make a rule that you'll leave the bathroom after five minutes even if you feel as if you have not completed evacuation.

How long does it usually take for a flare-up to subside?
If you carefully stick to the outlined routine, you should feel better in two to three days. If your problem persists beyond that, see your doctor.

What should I do if my internal hemorrhoids protrude?
If an internal hemorrhoid swells and protrudes, which occasionally happens, especially after a difficult bowel movement, it is important to reduce it as soon as possible. This is easy to accomplish. Place a drop of lubricating jelly on the tip of your finger and gently push the protrusion back into the rectum.

Why do hemorrhoids itch?
Internal hemorrhoids often leak mucus. This fluid makes the anal area moist and causes irritation and itching.

What can be done to stop the itching?
Keep the area clean and dry at all times. Cleanse yourself carefully after each bowel movement and pat yourself dry. It may also help to cleanse the area at other times during the day, using the procedure described on page 42. After drying yourself carefully, you can apply a gentle, nonperfumed talc or an astringent such as witch hazel or Balneol®. No matter how much you itch, never rub or scratch the area; this will only injure the sensitive tissues even further. If the problem persists,

your doctor can prescribe a medication to stop the itching.

Is bleeding a serious problem with hemorrhoids?
The bleeding that occurs with internal hemorrhoids is usually not serious. In most cases, it is intermittent and occurs only at the time of a bowel movement. Your doctor can inject the internal hemorrhoids with a solution that will stop the bleeding, but this is rarely necessary. Only a small amount of blood is lost, though it may appear to be much more to you when you see it on toilet tissue or in the bowl. The most important thing to remember about rectal bleeding is that no matter how much or how little occurs, it should always be brought to the attention of your doctor. Rectal bleeding can be the first sign of cancer.

A word of caution: If you do have bleeding hemorrhoids, stay away from aspirin and any products that contain it. Aspirin may increase the bleeding because of its effect on the time it takes for blood to clot.

Can any other serious problems result from hemorrhoids?
Other than local problems in the rectal and anal area, hemorrhoids rarely cause serious problems. They do not produce any systemic infection, nor do the blood clots that form in the veins travel to other parts of the body.

Is surgery always necessary to clear up the problem?
Absolutely not. Surgery is used only as a last resort. Most hemorrhoid problems can be handled either at home or in the doctor's office.

What can a doctor do in the office?
The common procedure is called ligation. This involves tying off the individual internal hemorrhoids with small rubber bands; these bands cut off all circulation to the hemorrhoidal veins, and, in a few days to a week, they will just drop off, usually during defecation. This rela-

tively minor office procedure is painless and highly effective. In some cases, your doctor may have to repeat the procedure several times. But once the veins fall off, they will never bother you again. (Ligation cannot be applied to external hemorrhoids.)

Can hemorrhoids be cured?

The only way hemorrhoids can be cured is if they are removed, either by surgery or ligation. If one of those procedures is not performed, your hemorrhoids will probably remain with you the rest of your life. This does not mean that they will always bother you. If you are careful about your diet, watch your weight, try to prevent constipation, and avoid excessive lifting, standing, and sitting, you may never have a problem with them.

5
INTESTINAL GAS

Probably one of the most embarrassing of all digestive complaints is gas. But as unpleasant and untimely as it can often be, the passage of a certain amount of gas is a normal part of the human condition. Hippocrates, the father of modern medicine, was the first to note that "passing gas is necessary to well-being" (most probably said in an attempt to placate his patients at the time). Even today's physicians cannot offer a way to prevent the passing of gas, but there are ways to reduce its amount—and to minimize your embarrassment.

What causes intestinal gas?

Intestinal gas comes primarily from two sources. Through swallowing, air enters the upper portion of the digestive tract; gas is also created in the lower intestine by the action of bacteria upon undigested food particles.

Exactly what *is* intestinal gas?

Intestinal gas is really a combination of many gases, including nitrogen, oxygen, carbon dioxide, hydrogen, and methane. Nitrogen, which is by far the largest component of intestinal gas, comes almost exclusively from swallowed air. Oxygen, although in much smaller amounts, also comes from swallowed air. Carbon dioxide, hydrogen, and methane are all produced in the large intestine as a by-product of digestion and fermentation. Fermentation occurs as a result of the activity of the intestinal bacteria upon undigested waste materials. As a whole, about 50 to 70 percent of all the gas in the digestive tract comes from swallowing; the rest comes from the activity of the intestinal bacterial flora.

How is air swallowed?
Air is constantly swallowed as you talk, eat, and drink; the faster you perform these actions, the more air is swallowed. If you drink from a straw you will take in an extra amount of air. Many foods and beverages have a high air content, including carbonated drinks, beer, ice cream, soufflés, and beaten eggs. Surprisingly, an apple is almost 20 percent air. You are often taking in a lot more air than you realize.

What happens to this air?
Much of it is released by belching, but some does make its way down through the digestive tract and out through the rectum.

Do some people swallow more air than others?
Yes, some people are chronic air swallowers. In addition to the normal amount taken in while talking, eating, and drinking, they unconsciously suck in an excessive amount. This is a habit they're not even aware of. When patients complain of excessive belching, and I explain to them what's causing it, they are as surprised as they are relieved.

Even if you're not a chronic air swallower, you'll tend to take in more air when you're nervous. This is probably because you're not relaxed enough at those times to eat, drink, or talk at your normal pace.

Some people actually force themselves to belch. Psychologically, they feel that belching will relieve some other kind of abdominal distress. Often, without even realizing it, they will swallow a gulp of air before they force a belch. Millions of dollars are spent each year on effervescent, over-the-counter medications, such as Alka-Seltzer®. Although these products may be beneficial to some people, the added carbonation may be detrimental to those who complain of excessive belching. These products act just like carbonated beverages and add to the amount of swallowed air. The more air that is swallowed, the more distention occurs in the esophagus and stomach. This results in even more belching, often with associated heartburn.

What can be done to relieve a belching problem?
In a healthy person, belching is caused by swallowing air. Therefore, in order to curtail it to some degree, the obvious answer is to stop swallowing so much air. If you have a problem with excessive belching and your doctor has ruled out organic causes, ask yourself the following questions: Do I drink a lot of carbonated beverages or beer? Do I often drink through a straw? Do I eat a lot of aerated foods? Am I a constant gum-chewer or candy-sucker? While you are talking, notice if you swallow air as you punctuate your sentences. If the answer to any or all of these questions is in the affirmative, you can control your belching by changing your habits.

Can excessive gas cause a feeling of distention and bloating after meals?
Many people think so, but it's not true. In a scientific study of this phenomenon, researchers infused excess air into a group of volunteer subjects just before a meal. They found no correlation between the amount of gas in the stomach and the sensation of bloatedness reported. The reason is that some people handle gas better than others. Those who don't handle it well most likely have a condition called irritable bowel syndrome (IBS) that they are not aware of. (IBS is discussed in greater detail in Chapter 11.)

Intestinal Gas

What causes excessive amounts of intestinal gas?
Most of the gas passed rectally comes from the activity of normal intestinal flora upon undigested foodstuffs. Therefore, the more undigested food present, the more gas produced. Fiber, by definition, is almost totally undigestible. Although a high-fiber diet can be a great help with constipation and bowel habits, it can cause an uncomfortable amount of gas.

But isn't eating fiber recommended?
Yes, but how *much* fiber you should eat depends on how well your system handles it. You'll have to find your own middle ground by trial and error. What may seem to be an excessive amount of fiber for one person is a perfectly acceptable amount for another. Certain high-fiber foods tend to cause more gas than others. Beans, celery, broccoli, raisins, cauliflower, carrots, onions, and cabbage are among the worst offenders. But if these foods are introduced gradually into the diet, the intestinal bacterial flora may adapt in such a way that these foods become less gas-producing. Again, this varies with the individual. Certain foods may be so gaseous to certain people that they must be avoided entirely.

A good way to tell which foods may have an adverse effect on you is to keep a record for several days of the foods you eat, the time you eat them, and the time you experience any gas problems. It usually takes less than six hours for a meal to reach the lower part of the colon. With that knowledge you'll be able to determine which foods are the worst offenders.

Why do people on a diet seem to have more gas?
The most obvious reason is that most reducing diets call for a large amount of high-fiber foods because these are usually lowest in calories. But another important reason is that people on a diet often use artificial sweeteners. These sweeteners are nondigestible carbohydrates that arrive in the lower bowel in their original

state. The bacteria act upon these sugars, and the resulting fermenting process produces more rectal gas.

If you are on a reducing diet and find excessive gas a problem, you should eliminate artificial sweeteners before you attempt to decrease your intake of high-fiber foods.

Are there any other causes of excessive gas?

Yes. Some people have a milk intolerance they may not be aware of. Milk, ice cream, and cheese require a special enzyme called lactase to be properly digested. In those who do not have enough of this enzyme, products containing milk sugar (lactose) arrive in the colon undigested and undergo fermentation by colonic bacteria; this causes an excessive amount of gas. If you notice, after keeping a list of the foods you eat, that your gas problem occurs about six hours after consuming milk or milk products, discuss it with your doctor. He or she may arrange for you to take a test to determine whether you are deficient in lactase.

Are there any medications that might help?

There are medications on the market containing simethicone, an ingredient used to treat excessive gas problems. These products supposedly act by combining the smaller gas bubbles into larger ones so that they can be more easily expelled. The beneficial effects of simethicone, in tablet or liquid form (Di-Gel®, Mylicon®), have not been well documented in clinical trials. The same applies to charcoal, widely touted as an absorber of gas. Charcoal is useful in the absorption of ingested poisons but, in my experience, neither charcoal, simethicone, nor any medication is as effective in relieving the symptoms of gaseousness as is diet modification.

What causes gas pains?

Gas pains can be distressful, but they are usually not a serious problem. What happens is that gas is trapped along the digestive tract as peristalsis (rhythmic, involuntary contractions of the alimentary canal) occurs. As the muscular contractions move the gas along, the pain goes away.

Can gas pains ever be serious?
Yes, if they are caused by an obstruction that blocks the normal passage of gas out of the rectum. Because of this possibility, you should see your doctor if gas pains persist for any length of time or if you develop any other symptoms with them. Another reason to check with your doctor is that sharp abdominal pains could be caused by something other than gas. A persistent problem with abdominal pain merits medical attention.

Can anything be done to relieve gas pains?
Gas pains as such are usually transient, passing spontaneously within several seconds. Any pain or discomfort in the abdomen that does not resolve quickly may relate to a more serious condition and should be brought to the attention of a physician.

Distention or bloating, rather than pain, is the sensation many people feel after meals, and is often attributed to excess gas. But excess gas is only part of the explanation. The abdominal muscles that overlie the intestines help to contain intestinal gas and limit distention of the overlying abdominal wall in the same way a corset or tight clothing might prevent abdominal "bulge." With increasing age, decreasing physical activity, or both, the abdominal muscles become lax and less effective in this regard. Doing proper exercises, such as sit-ups, will tighten these muscles and keep them well developed. This helps to prevent excessive distention and reduce the bloated feeling after meals.

In some people, however, weakness in the abdominal muscles may be the result of an abdominal hernia. This condition may occur spontaneously or after surgery of the abdomen, and is the result of separation of the abdominal muscles themselves. This type of hernia, although not a serious problem, often contributes to the sensation of distention. An abdominal hernia cannot be corrected with exercise; in rare circumstances, it requires surgery.

Is there a link between rectal gas and cancer?
Studies have found that some people with bowel cancer produce a larger than normal amount of methane in their rectal gas, and this overproduction may be the result of an early environmental factor. These findings may help us to eventually identify some of those people who have a greater tendency to develop bowel cancer and to prevent it from occurring.

6

GALLSTONES

Cholecystitis is a medical term for what many people simply call gallbladder trouble. The term refers to inflammation of the gallbladder, which can bring on a sudden attack of excruciating pain accompanied by nausea, vomiting, and fever.

Cholelithiasis, the medical term for gallstones, is so often associated with cholecystitis that they are frequently thought of as one problem—which they are more than 90 percent of the time. Gallstones may obstruct the free flow of bile; the resulting inflammation causes gallbladder disease with its symptoms and problems. But the two conditions do not always occur together. Gallstones can be present and not cause any symptoms at all, while a gallbladder attack can occur in the absence of gallstones.

Gallbladder disease, like many other digestive disorders, is surrounded by myths and misinformation, leaving people unsure of what causes the disease and what medically sound treatments are available. This chapter explores every aspect of the disease, from the obsolete "low-fat gallbladder diet" to the newest medical discoveries in the diagnosis and treatment of the problem.

What is the function of the gallbladder?
The gallbladder, a small, pear-shaped, hollow organ, lies in the abdomen, adjacent to and underneath the liver. Its purpose is to remove the excess water from bile, concentrate the bile, and then store it until it is needed in the small intestine to digest dietary fats. During meals, the duodenum signals the need for bile. The gallbladder then releases about four tablespoons of bile and sends it into the duodenum via the bile duct.

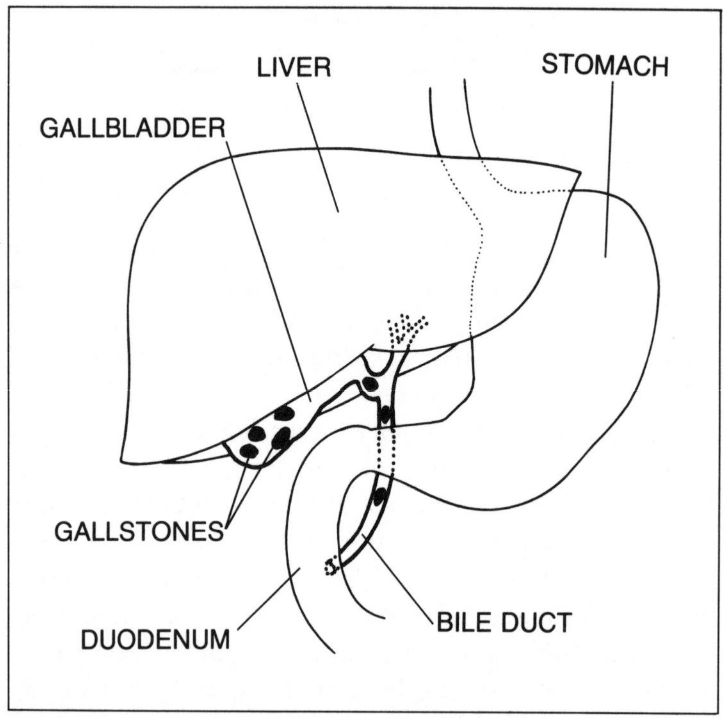

Gallstones in the gallbladder and lodged in the duct.

Is the gallbladder a vital organ?
No, it is not. People who have their gallbladders removed can live totally normal lives.

What are the symptoms of a gallbladder attack?
The symptoms of an acute attack are dramatic. Usually, there is a constant severe pain in the right upper

portion of the abdomen just under the ribs that may radiate to the right shoulder. The pain of a gallbladder attack, one of the most severe pains one can experience, may be accompanied by nausea, vomiting, and fever.

What causes the gallbladder to malfunction and an attack to occur?

A gallbladder attack is usually caused by gallstones and the inflammation that results from them. One or more gallstones become lodged in the duct that carries bile to the duodenum. When this duct is blocked, the normal flow of bile is interrupted and the gallbladder becomes acutely inflamed.

Can you have a gallbladder attack without having any stones?

This *can* happen, but usually it means that stones have blocked the duct at some time in the past. Although they may have dissolved or passed out of the body on their own, the obstructive damage they caused may have produced inflammation, infection, and pus within the gallbladder itself. This diseased gallbladder causes the current symptoms.

Is an infection in the gallbladder serious?

Absolutely, especially in people over fifty years of age. If this condition, known as empyema of the gallbladder, is not diagnosed and treated early, it can even be fatal.

What are the symptoms of empyema of the gallbladder?

Unfortunately, the symptoms vary in degree and intensity, making this a difficult disease to diagnose. The major symptom is abdominal pain, which may or may not be severe and which may subside on its own while the infection is rapidly escalating. There may be jaundice (yellowing of the skin), loss of appetite, and a general feeling of malaise; or there may be a mass in the upper right side of the abdomen that the physician can feel upon examination. Notice, I said "*may* be." Sometimes the symptoms are so vague that in the early stage people are not even aware they are ill.

Are there any reliable diagnostic tests for empyema of the gallbladder?

The one test that seems to confirm the presence of this disease is a blood culture. The blood is treated in the laboratory in such a way that the bacteria will grow in a test tube if an infection is present.

How is empyema of the gallbladder treated?

The big danger with empyema is that the infection will spread throughout the body once the pus spills out of the gallbladder and into the bloodstream. Most doctors will insert a drain into the gallbladder or duct to drain off the pus and, at the same time, treat the person rigorously with antibiotics. If the diagnosis is made early enough, this treatment is successful.

How is the presence of gallstones diagnosed?

If your doctor suspects you might have gallstones, he or she will order some specific tests, such as a cholecystogram or ultrasound. An oral cholecystogram, or OCG, begins the evening before the examination. The person swallows two to four tablets composed of a dye that is absorbed from the stomach and then excreted by the liver into the gallbladder. On x-ray film, stones will appear as dark spots in the gallbladder. However, the stones can be so small that they will not be visible by means of this type of x-ray examination.

A recently developed test is even more reliable. Called ultrasonography, it does not utilize x-rays. High-pitched, high-frequency sound waves are transmitted through the body over the area of the gallbladder. A sound-producing device called a transducer is placed over the skin and moved from one location to another. If there is a stone in the gallbladder, an echo is generated in that area, which is then transmitted through a computer onto a video screen. A written record is made from that image.

Your doctor may also order a special test called endoscopic retrograde cholangiography and pancreatography, or ERCP for short. This test is more complex than those already described, and must be done by a specialist. An endoscope (a flexible fiberoptic tube) is passed into the mouth and stomach until the opening of the bile duct, in the intestine, is seen. A catheter (or small tube) is passed through the endoscope and threaded into the bile duct. Dye can then be injected into the duct; x-ray films are then taken to show the bile ducts and any stones that may be present. This test also makes the pancreatic duct visible, enabling your doctor to determine whether disease is present there.

If small stones are found within the bile duct, they can sometimes be removed through the endoscope, making surgery unnecessary. However, this is usually reserved for people who have already undergone gallbladder surgery and who have stones within the bile duct.

How common are gallstones?

Anyone can develop gallstones, but they become far more prevalent with advancing age. It is estimated that 30 percent of women and 15 percent of men over the age of seventy have gallstones. In fact, gallbladder surgery is the most common type of abdominal surgery performed in this age group.

Does alcohol use predispose a person to developing gallstones?

No. Drinkers and nondrinkers seem to develop gallstones with equal frequency.

Are all gallstones the same?

No. There are two different kinds of gallstones. Bilirubin stones are formed by the continuous premature breakdown of red blood cells. These stones are rare and occur only in people who have certain hemolytic blood diseases, such as sickle cell anemia.

The most common kind of gallstone is the cholesterol stone, which, as the name indicates, is primarily made up of concentrated cholesterol. It is estimated that from 85 to 90 percent of all gallstones removed surgically are cholesterol stones. (The development of these stones, incidentally, is *not* related to a high blood cholesterol level.)

What causes cholesterol gallstones to form?
Many experts think that people who develop cholesterol gallstones have abnormal bile, known as "lithogenic bile." The ratio of bile acids to cholesterol is disturbed; there is a relative excess of cholesterol, which leads to the formation of stones.

What do cholesterol gallstones look like?
Cholesterol stones resemble little beach pebbles. They range in size from less than an inch to more than two inches in diameter. During surgery, the surgeon may discover thirty or more stones, but only one stone may be wedged in the bile duct and therefore be responsible for the gallbladder attack.

Does everyone with gallstones experience painful symptoms?
No. Many people have gallstones and never know it until the stones are discovered incidentally during a diagnostic evaluation for an unrelated problem. We call these phenomena "silent" stones. They can float freely within the gallbladder and remain that way throughout a person's life. Problems arise only when one or more of them lodge in the bile duct.

If silent stones are discovered, should surgery be undertaken before a problem develops?
No. The percentage of people with silent stones who eventually develop a problem requiring surgical intervention is quite small. But there is one possible excep-

tion: If silent stones are discovered in a person with diabetes, surgery sometimes is performed even if there are no symptoms. (This is because diabetics are particularly prone to serious complications if a stone becomes lodged in the duct.) However, this practice may be changing. Recent evidence suggests that even diabetics with silent stones may forgo surgery if they are free of symptoms.

Can gallstones be a serious problem?
Yes. Once a stone is lodged in the duct, many complications can occur. Inflammation and infection are the most common, but a stone may erode its way through the lining of the gallbladder and cause perforations, fistulas, or peritonitis. It may also cause severe liver problems by blocking bile secretion.

Can gallstones go away by themselves?
Sometimes a stone will dissolve or pass out of the duct into the duodenum. Although the acute problem has passed, it is likely that the person will soon develop another stone that will cause problems.

How are gallstones treated?
The most common method to deal with gallstones in the duct is to surgically remove the gallbladder and, at the same time, explore the bile duct to make sure that all stones lodged in it are removed. The bile duct itself is not extracted because it is needed to transport bile from the liver into the small intestine.

Are any drug treatments available?
Recently, there have been trials of new drugs called cheno- and ursodeoxycholic acid. These bile acids are normally secreted in human bile and can dissolve cholesterol gallstones. Their use is currently reserved for certain people with silent stones or for people who are too weak or debilitated to undergo surgery. With these medications, it takes many months for the stones to

dissolve; the drugs must then be continued indefinitely to prevent the stones from forming again. Another drawback is that many patients on these medications develop diarrhea. It is hoped that eventually bile acids will be developed to a point where surgery can be avoided.

Can gallstones recur after the gallbladder has been removed?

Yes. A stone may have been lodged in the liver and only dropped down into the bile duct after the gallbladder has been taken out. But this is quite uncommon.

If that happened, would a person be advised to have surgery again?

Not necessarily. Because the gallbladder has already been removed, all that is necessary may be removal of the offending stone from the bile duct. ERCP, the procedure I described earlier, is helpful in this case. The stone or stones can sometimes be extracted directly through the endoscope, eliminating the need for repeat surgery.

Can someone develop a stone many years after surgery?

Yes, though this is unusual. No one really knows why, but a stone may form within the liver and drop down into the bile duct years after surgery. When this happens, the symptoms are exactly the same as with the initial attack. More surgery may be necessary.

Is there such a thing as a "gallbladder diet"?

Years ago it was thought that because the gallbladder is involved in the digestion of fats, limiting the amount of fat in the diet could relieve the symptoms or even help to avoid a future attack. But there is no evidence to confirm this theory. The amount of fat in the diet does not seem to be related either to the formation of the stones or to their obstruction of the bile duct.

If you have had gallstones, are you more likely to develop cancer in that area?
Most authorities do not believe the presence of stones makes malignant change in the liver more likely.

7

ULCERS

Ulcer disease is sometimes talked about in a jocular way. Comedians joke about their mothers-in-law giving them ulcers, and cartoons depict the ulcer as a success symbol of the aggressive, hard-driven businessman. But for the more than 20 million Americans who suffer from ulcer disease, it is no laughing matter. Ulcers can be painful and even disabling. They often appear as life-threatening emergencies. Even in these medically sophisticated times, twelve thousand Americans each year die from complications of ulcer disease.

In this chapter, many of the myths about ulcer disease will be dispelled. Milk is not always the "cure all" it was once thought to be, nor is the "traditional" ulcer diet of much help in treating the disease. It is not only the hard-driven businessman who gets an ulcer but also the part-time worker, the hospital volunteer, and even the person who is fully retired.

The real facts about ulcer disease will be conveyed, including exactly what this illness is, how ulcer craters form, and what can be done to treat ulcers successfully. This chapter will also alert you to the complications that can occur and how they can be avoided.

What is an ulcer?

Though you may think of an ulcer as a problem confined to the digestive tract, this is not the case. An ulcer can appear on almost any organ of the body, including the skin, mouth, leg, or even in the eye. By definition, an ulcer is any break in the mucous membrane or skin that results in an open sore. A variety of medical conditions can lead to ulcers on various parts of the body. For the purpose of this discussion, which is confined to problems of the digestive tract, only the two most common kinds of ulcers found in the digestive organs will be described: gastric ulcers and duodenal ulcers.

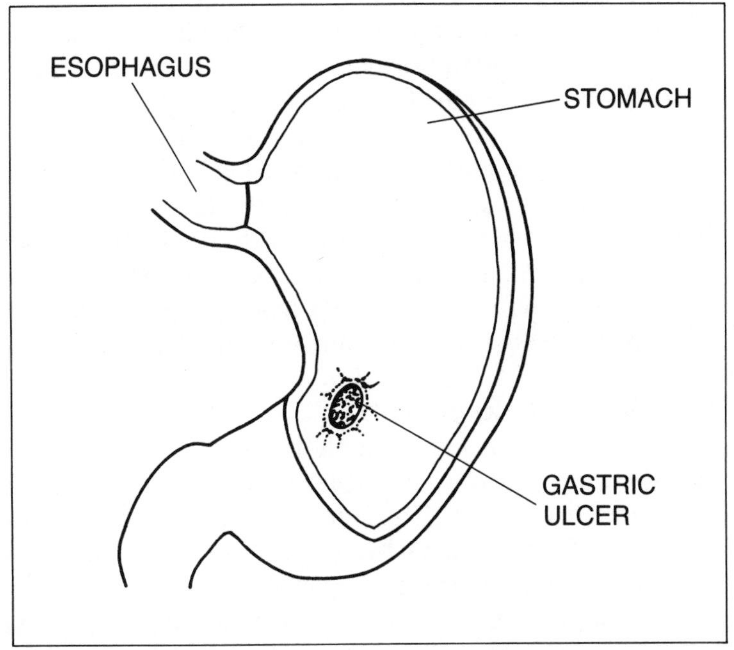

The stomach, with gastric ulcer.

What is a gastric ulcer?

Any ulcer appearing within the stomach is called a gastric ulcer. (If such ulcers occur directly opposite each other on the walls of the stomach, they are known as kissing ulcers.)

What is a duodenal ulcer?
Any ulcer appearing within the duodenum, or just at the point where the stomach and duodenum join together, is called a duodenal ulcer.

What is a malignant ulcer?
A malignant ulcer is a cancerous growth that may take on the appearance of a benign peptic ulcer. Malignant ulcers occur more commonly in the stomach than in the duodenum; in contrast, benign ulcers occur more commonly in the duodenum.

What do ulcers look like?
Most of the lining of the digestive tract and its organs is pink. But an ulcer, if it is not bleeding or has not recently bled, is white or pale yellow. This color results from fibrin, a protein that develops during the healing process and accumulates within the ulcer crater. If the ulcer is actively bleeding, it appears as an open crater with blood oozing out or, at a later stage, as a crater with a blood clot in the center. Although ulcers can vary in size from microscopic to about an inch or more in diameter, their size has little bearing on the amount of pain they cause or the complications that can occur.

Why are ulcers so painful?
Contrary to what many people think, the ulcer itself is not the only cause of pain. Many factors play a role in producing ulcer pain. These include excessive stomach acidity and irritation of the entire stomach lining. Ulcers may heal, but the pain persists if the excess acid within the stomach is not diminished with continued treatment. On the other hand, with prompt medical treatment, the pain may be relieved even before the ulcer completely heals.

What are the signs that an ulcer is present?
If you have pain in the area right under your breastbone that seems to recur frequently, especially when your stomach is empty, there is a chance you may have

an ulcer. If you have such a chronic symptom, talk to your physician about it.

How can a doctor confirm the presence of an ulcer?
If you develop suspicious symptoms, your doctor will order some diagnostic tests. Initially, an x-ray film called an upper gastrointestinal series (or UGI) is done. The person swallows a barium solution that makes it possible for the radiologist to see the conformation of the stomach and small intestine. If a duodenal ulcer is found, the person returns to the referring doctor and a medical regimen is prescribed. If a gastric ulcer is discovered, the doctor will probably investigate further because what appears on x-ray film examination to be a benign peptic ulcer may actually be a cancerous growth. In fact, this is the case about 2 to 3 percent of the time.

How can a benign ulcer be distinguished from a malignant growth?
The test for this is called endoscopy. While the person is under mild sedation, the doctor passes a gastroscope gently down the mouth and into the stomach and small intestine. The gastroscope is a flexible fiberoptic tube with a light at one end and a camera at the other. It allows the doctor to see the ulcer crater and photograph it for further study. A channel within the gastroscope allows the passage of a biopsy forceps so that tissue samples may be obtained. This procedure and a UGI x-ray film series are often performed on a regular basis in people with gastric ulcers.

How is a stomach malignancy treated?
Treatment may include surgery, chemotherapy, and radiation. The type of treatment chosen depends on the type of malignancy present.

Can gastric ulcers become malignant?
No. Originally it was thought they could because a certain percentage turned out to be malignant. But now,

with more sophisticated diagnostic tools, it has become evident that the problem was not that gastric ulcers were becoming cancerous, but that the malignancy was not recognized as such in the first place.

How common are gastric and duodenal ulcers?
Duodenal ulcers are the most common of all digestive tract ulcers; more than 76 percent of ulcer craters are found in the upper portion of the duodenum. Approximately 24 percent of digestive tract ulcers are in the stomach.

Are some people more likely than others to develop gastric ulcers?
Yes. More than 80 percent of all duodenal ulcers and 70 percent of all gastric ulcers occur in men, although we do not know why this is the case. Recent statistics show a rise in the incidence of gastric ulcers in women, but men still have the lead.

Can anyone develop a peptic ulcer?
Yes, peptic ulcers are found in people of every age group and of every nationality. There is a slightly increased risk of ulcer disease in people with blood type O. Approximately one-third of people with peptic ulcers are over sixty years old—a rise of more than 36 percent in the last ten years.

Is ulcer disease the same in people over fifty as it is in younger people?
No, it is quite different. The symptoms are less definitive, the beginnings of the ulcer are harder to pinpoint, and the treatment is more difficult. For reasons we don't completely understand, older people do not respond as well to medication as do their younger counterparts.

Are ulcers inherited?
The hereditary factor involved in ulcers is still not understood. However, it is known that if one of your close

relatives has had ulcers, you are more likely to develop the disease than if no relative has had ulcers.

What are the symptoms of a duodenal ulcer?
A particular pain pattern, known as "pain-food-relief," is associated with all peptic ulcers, duodenal or gastric. Generally, a gnawing pain occurs just under the breastbone when the stomach is empty. It is relieved within minutes of taking an antacid or eating something. Duodenal ulcer pain has a tendency to recur in the middle of the night, around 2 A.M., when stomach acid levels peak.

What causes a duodenal ulcer?
A duodenal ulcer is caused by an overproduction of stomach acid pouring into the duodenum, causing a superficial erosion that develops into an ulcer.

Can a duodenal ulcer become malignant?
Duodenal cancer is rare. When it does occur, neither its pain pattern nor its appearance on an x-ray film resembles that of a duodenal ulcer.

What are the symptoms of a gastric ulcer?
The symptoms of gastric ulcer are the same as those of a duodenal ulcer only when the underlying mechanism is excess acidity within the stomach.

Aren't gastric ulcers, like duodenal ulcers, caused by excessive stomach acid?
Not always. With duodenal ulcers, stomach acidity is always elevated. However, measurements of stomach acid levels in people with gastric ulcers have shown that not all of these people have excess acid. In fact, many people with gastric ulcers have less stomach acid than would be expected.

Then what does cause a gastric ulcer?
Although the origins of gastric ulcers are not fully understood, we do know some of the triggering factors.

An acute gastric ulcer can be a situational problem produced by the body's reaction to a stress situation, such as the loss of a loved one or retirement. Some drugs can cause an ulcer in the stomach, especially aspirin and many of the antiarthritic medications on the market, such as ibuprofen (Advil®, Rufen®, Nuprin®, and Motrin®) and indomethacin (Indocin®). (See p. 71.) Certain physical illnesses, such as a severe infection, pulmonary disease, a heart attack, uremia, severe burn injuries, hormonal problems, or even certain types of major surgery, can trigger a gastric ulcer.

But these acute kinds of gastric ulcers are one-time problems. Once healed, they are unlikely to recur, unless the offending agent or problem returns.

Are ulcers curable?

Both gastric and duodenal ulcers can be treated medically (see p. 71), but the underlying hyperacidity within the stomach persists unless ulcer surgery is performed. Fortunately, however, ulcer surgery can generally be avoided. With modern anti-ulcer medication, gastric hyperacidity can be suppressed and symptomatic flare-ups can be kept to a minimum.

Are emotions a factor in ulcer disease?

Many jokes are made about the typical "ulcer personality." The profile is of a person who is aggressive and driven toward achievement and success. This theory lacks hard evidence, however. Psychological studies have found that stress can exacerbate already existing ulcer disease, but so far no direct link between personality type and ulcer disease has been discovered.

Can psychotherapy help in the treatment of ulcer disease?

Sometimes, especially in the case of acute stress ulcers (those related to physical or emotional stress), psychotherapy may help. If a person is having problems coping with life, obtaining professional help is certainly to be recommended, whether ulcer disease is present or not. But so far, evidence does not conclusively indicate that ulcer disease is either caused by emotional problems or helped by the treatment of these problems.

What complications can occur with ulcer disease?
There are four main complications: bleeding, obstruction, perforation, and intractability. Each one can precipitate a life-threatening emergency. When one or more of these complications arise, surgery may be necessary. Let's discuss each complication separately.

Bleeding Ulcers can bleed and often do. An ulcer can penetrate the mucous membrane lining of the stomach or duodenum until it actually erodes into a small blood vessel. When an artery is eroded, bleeding is profuse and dangerous. Large amounts of blood can be lost quickly. Within several hours, a person can lose half of all circulating blood, resulting in shock or even death. With arterial bleeding, a person will either vomit up red blood or material that looks like coffee grounds but is actually blood mixed with some of the digestive enzymes. The only indication that a smaller blood vessel is bleeding may be the passage of black, tarry stools. If you have an ulcer, it is wise to check the color of your stools daily. A bleeding ulcer requires immediate treatment. The bleeding must be stopped, and the person must be monitored for signs of further complications.

Obstruction Obstruction can result from scarring and swelling that occur each time an ulcer reappears. When there is obstruction, the opening (pylorus) between the stomach and the duodenum narrows, preventing the passage of stomach contents down the digestive tract. Unremitting abdominal pain and distention with nausea and vomiting may be initial indications of obstruction. Sometimes an ulcer is first discovered when the persistent vomiting resulting from an acute obstruction necessitates surgical intervention.

Perforation Acute perforation accounts for the greatest number of deaths from ulcer disease. When an ulcer perforates, it erodes all the way through the mucosal wall of the stomach or the duodenum, dumping the contents into the abdominal cavity. (It may even invade other organs, such as the gallbladder or pancreas.) Acute perforation often leads to peritonitis, infection of the abdominal lining. If treatment is not initiated immediately, death can follow.

Intractability An intractable ulcer does not respond to ordinary conservative medical treatment or, if it does, reappears rapidly after the healing process has taken place. If all other measures fail, surgical intervention is necessary.

How should I treat an ulcer to prevent these problems?

There is no guarantee that proper treatment will prevent complications. Some patients with acute episodes of bleeding or obstruction have closely followed their physician's treatment recommendations. Other patients have ulcers for years and yet never develop any complications at all. It makes sense that the faster an ulcer crater heals, the fewer problems it causes.

What is the best treatment for the ulcer crater?

First of all, it is important to clear up some myths about treating an ulcer. The first myth involves milk. It has long been believed that people with ulcers should drink large quantities of milk and eat primarily milk products. But we now know that milk is not a panacea. The calcium in milk can actually cause the release of more stomach acid. Although it seems to coat the stomach, milk does little to neutralize stomach acid. An occasional glass of milk is fine, but it is not recommended in large quantities or as the sole treatment for an ulcer.

Another popular myth about ulcers is that spicy foods aggravate an ulcer or prevent healing. This is just not true. A diet consisting exclusively of highly seasoned food three times a day is not a good idea, but there is no reason why you cannot occasionally eat spicy foods. If you have an ulcer, let it be the best judge. It will let you know which foods bother you and which foods do not.

Some people believe that those with ulcer disease should limit their activities. But, this is not necessary unless there has been associated bleeding or anemia. If you are not having an acute flare-up that is causing pain, go on with your normal routine.

Some of the real rights and wrongs of ulcer treatment may not be as well known as the myths. Alcohol, cola, coffee, aspirin, and smoking are all potent acid producers and/or irritants in ulcer disease. If you are a smoker and you have an ulcer, this is the best reason in the world for you to stop smoking. Coffee, whether or not it contains caffeine, may adversely affect ulcer disease. Alcohol is no less an offender, even when diluted with water; the alcohol still reaches the stomach and duodenum and exacerbates ulcer symptoms. Aspirin should also be avoided. Even one aspirin can cause a small ulcer. The same applies to the antiarthritic medications referred to as NSAIDs or nonsteroidal anti-inflammatory drugs. Examples of these are Motrin®, Naprosyn®, and Clinoril®, but there are many others. One of these medications, ibuprofen, is now available without a prescription (Advil® and Nuprin®). Although ibuprofen is available only in a relatively safe, low dose, it is still an NSAID and as such can exacerbate ulcer disease. Ibuprofen should be used with caution, if at all, by anyone with an ulcer history. Acetaminophen (Tylenol®, Datril®) is an excellent aspirin substitute, as it will not exacerbate ulcer symptoms.

What is the best treatment for an ulcer?

The best treatment for a benign peptic ulcer is a combination of liquid antacids and one of three medications. The first two, cimetidine (Tagamet®) and ranitidine (Zantac®), are referred to as H_2-receptor antagonists. They heal ulcers by blocking the release of acid within the stomach. The third medication, sucralfate (Carafate®), facilitates ulcer healing by reinforcing the protective surface lining of the stomach, a process referred to as cytoprotection.

Your physician may prescribe other medications according to your specific needs. But these medications are effective in healing and controlling symptoms of peptic ulcer disease.

Are all antacids the same?

No. The chemical composition of all the popular antacids differs. One may be better for you than another, depending on your particular needs.

How do I know which antacid to take?

The best thing to do is consult your physician, as he or she is most knowledgeable about your medical condition. There are, however, a few general considerations about the different antacids that may help you and your doctor make the right decision. Some antacids have a higher salt content than others, so they could pose a problem if you are on a sodium-restricted diet. Many of the newer antacids currently on the market have a low salt content and can be recommended to people who are watching their salt intake.

All antacids will affect the absorption of other medications you may be taking, such as antibiotics and tranquilizers. If your doctor prescribes a medication for you, make sure you tell him or her about any antacids you are taking.

Antacids containing aluminum hydroxide and/or calcium carbonate tend to cause constipation, so avoid them if constipation is a problem for you. Antacids containing magnesium hydroxide tend to cause diarrhea if taken in large quantities. Some antacids contain both aluminum and magnesium, and do not adversely affect bowel functioning.

All antacids containing calcium or magnesium have the potential of causing kidney problems if taken over an extended period of time, especially if there is an underlying kidney disorder. Although antacids may be sold over the counter, they are drugs that may alter bodily functions.

What is the best way to take an antacid?

All antacids should be taken in liquid form. They work faster and more efficiently this way. If you find liquid antacids less palatable than tablets, try refrigerating them first.

If you have an ulcer, the time antacids are taken influences their effectiveness. Antacids should be taken when the stomach is empty, about one hour after meals. If you wake up in the middle of the night with ulcer pain, this is a good time to take antacids. The acid level in the stomach peaks at about 2 A.M. or 3 A.M. and often creates a great deal of pain and discomfort.

Are large doses of antacids good for everyone?

The long-term effects of antacid use include depletion of certain important nutrients. Therefore, antacids should not be used indiscriminately or for prolonged periods of time. Aluminum antacids reduce serum phosphorus levels. Combination aluminum and magnesium antacids reduce serum proteins and calcium. Older people who may already have depleted levels of vitamin D as well as osteoporosis should not risk further loss of vitamin D or calcium. The sodium content of antacids varies, and excessive use of these products can result in heart failure and fluid retention in those older individuals with underlying coronary or kidney disease. Antacids are safe when used properly, in moderation, and only when necessary.

How long does it take an ulcer to heal?

Once therapy is started, an ulcer will usually heal within four weeks. It is a good idea, however, for patients to stay on their medication for a full four to six weeks. This is to ensure that acid levels are kept at a minimum after healing has occurred.

8
HEARTBURN

Heartburn. The mere mention of the word brings to mind the uncomfortable burning sensation that suddenly makes you feel as though your whole chest is on fire.

Everyone has had heartburn. For some people it's only an occasional occurrence that goes away immediately when they drink a glass of milk or take an antacid. But for others, heartburn is an unavoidable fact of daily life.

The following pages explain the causes of heartburn and describe some simple tricks to help you deal effectively with this condition.

What is heartburn and what causes it?

Heartburn is not a medical term. The scientific term for that painful burning sensation in the middle of your chest is reflux esophagitis.

A sphincter called the lower esophageal sphincter separates the esophagus from the stomach at the level of the diaphragm. This sphincter acts something like a door, opening to allow food to go from the esophagus to the stomach and then, under normal circumstances,

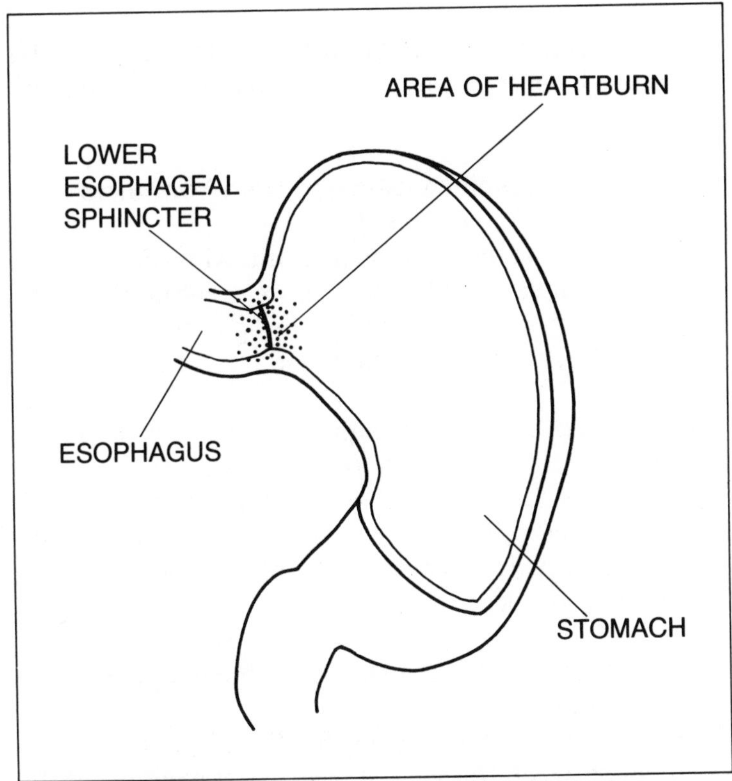

The lower esophageal sphincter, which separates the esophagus from the stomach.

closing immediately to prevent the contents of the stomach from returning back up into the esophagus. People who have chronic reflux esophagitis have a weakness in this sphincter that prevents the sphincter from doing its job properly. Because of the weakness, some of the stomach contents find their way back up into the esophagus, causing the hydrochloric acid in the stomach to irritate the esophageal walls.

How can I tell if I have reflux esophagitis?

The symptom of reflux esophagitis is a frequent burning sensation in the chest that has a particular pattern. It usually occurs right after eating and is aggravated by lying down after meals. Occasionally, if some of the intestinal bile is regurgitated into the esophagus, there may be a bitter taste in the mouth. The pain may be moderate to severe, but it will always be immediately relieved by an antacid.

How can I differentiate between pain from esophagitis and pain from an ulcer?

An ulcer will usually produce a gnawing pain in the upper abdomen. It will occur before mealtime and will be relieved by eating. Besides being painful during the day, an ulcer may also wake a person in the middle of the night.

The pain of reflux esophagitis is clearly "heartburn." It will be felt in the middle of the chest right near the heart. After you have lived with it for a while, you will realize that certain positions and habits will aggravate it. The burning pain is most acute after meals—just the opposite of an ulcer symptom.

Why can an attack of reflux esophagitis feel like a heart attack?

Sometimes it is difficult even for the physician to distinguish between the pain of reflux esophagitis and the pain of angina pectoris, or a heart attack, especially if the esophagitis is severe. This is because these problems stimulate the same nerve pathways. As a rule, a reflux esophagitis attack will be relieved as soon as you take an antacid. If the pain persists despite the use of antacids, or if the pain is different in character or distribution from prior episodes, call your doctor. A physical examination and electrocardiogram may be the only ways to exclude a coronary source of pain.

Is reflux esophagitis serious?

Usually it is not. Millions of people suffer from the symptoms without developing any ill effects except, of course, for the discomfort of the symptoms themselves. But if chronic esophagitis is not treated properly, it *can* evolve into a more serious problem. The constant back flow of acid can cause ulcers in the esophagus because those delicate tissues are susceptible to damage by stomach acid and bile. This can lead to scarring and then narrowing of the esophagus, which can make swallowing food difficult or even impossible.

Can reflux esophagitis cause cancer?

In most cases it does not. But if it is allowed to continue for many years without proper treatment, the constant irritation to the esophagus can cause malignant changes in the tissues.

What should I do when I get an attack?

If you feel that familiar burning sensation, take a liquid antacid immediately. The liquid works much better and faster than the tablets. It is sometimes unpleasant to drink, but if it is refrigerated, it becomes more palatable.

Stand up or walk around. Do not lie down.

Loosen all tight clothing, especially girdles, belts, and tight-fitting pants.

You should feel relief instantly. If the pain continues or if you develop any other symptoms with it, call your doctor.

Is Gaviscon®, which I can purchase over the counter, an antacid?

Gaviscon® contains both antacids and an ingredient called alginic acid. It is sometimes useful as part of the treatment of reflux esophagitis.

When swallowed, alginic acid dissolves into a foamy, viscous solution called sodium alginate. The foam forms a thick surface layer or "raft" that floats on top of the stomach contents and acts as a physical barrier to reflux.

When is the best time to take compounds such as Gaviscon®?

The main purpose of alginic acid is to prevent the reflux of stomach contents into the esophagus. This occurs, most frequently, immediately following meals. For that reason, Gaviscon® and similar medications are best taken on a full stomach right after eating. This is in contrast to pure antacids, which do not prevent reflux but are potent neutralizers of acid; for that reason, pure antacids such as Maalox® and Mylanta® are best taken at least one hour after meals to neutralize the acid present in the stomach after it has emptied.

Can anything else help if these products are not available?

Believe it or not, water works well. If you drink a glass of cold water, it will wash the acid from the surface of the esophagus back into your stomach. It may not be as effective as an antacid, but it does help.

Should I try a glass of milk?

Milk has long been thought of as a natural antacid. Initially, it may act as a buffer, but its fat content may actually relax the lower esophageal sphincter and contribute to further reflux. Also, its high calcium content may stimulate increased acid production.

Are there any prescription medications available to treat reflux esophagitis?

Yes. Those medications that reduce stomach acid secretion also ameliorate the symptoms of reflux esophagitis. These medications are referred to as H_2-receptor antagonists (Tagamet®, Zantac®), and are available by prescription only.

Another medication useful in the treatment of severe reflux esophagitis is metoclopramide (Reglan®), also available by prescription only. It is neither an antacid nor an H_2-receptor antagonist, and it has no direct effect on stomach acid. It is referred to as a dopamine-receptor blocker. This medication works by causing the lower esophageal sphincter muscle to contract, thus preventing reflux of stomach contents and acid into the esophagus. At the same time, it stimulates the emptying of food from the stomach. However, metoclopramide infrequently induces parkinsonism and, rarely, tardive dyskinesia, a form of involuntary movement. A more common side effect is fatigue. For these reasons, its use is reserved for severe symptoms not relieved by other means. It should not be used for prolonged periods of time, and its use in people over fifty is limited.

If an attack of heartburn makes me feel like belching, would it help to take something to bring up the trapped air?

No. This is a common misconception. Many people think heartburn can be relieved by an over-the-counter effervescent product that makes them belch. The idea is that this will relieve some of the pressure and discomfort. In fact, it really aggravates the problem. When you belch, you may bring up some stomach acid with the air. Obviously, this is not what you want. Another drawback is that some of the effervescent products contain aspirin, and aspirin irritates the mucosal lining even further.

Carbonated beverages will also make you belch, so stay away from them when you have heartburn.

Can anyone develop reflux esophagitis?

Yes, but certain people are more predisposed than others to developing reflux esophagitis. People who are obese, chronically constipated, and who do a lot of straining, lifting, and carrying increase the amount of intraabdominal pressure and are therefore more likely to develop the problem.

How can I control the symptoms of reflux esophagitis?

First of all, if you are overweight, you should go on a reducing diet. Excess weight increases intraabdominal pressure, which aggravates the reflux. Constipation and straining with defecation also increase intraabdominal pressure. Follow the regimen outlined in the chapter on constipation, and add fiber and fluids to your diet.

Avoid wearing tight clothing, including girdles, belts, tight pants, and anything else that puts pressure on the waist area. Avoid bending at the waist. If you have to bend down to pick something up, bend at the knees. This will also help prevent back problems. Never lie down after you eat. Wait at least two to three hours after eating before you take a nap or go to sleep for the night.

Do not prop yourself up in bed while reading or watching television. Without realizing it, you are bending and adding pressure to the problem area. Elevate the head of your bed by placing six-inch blocks under each of the two leg posts. This is the ideal position to sleep in when you have the symptoms of reflux esophagitis (whether or not you have hiatus hernia). It allows gravity to help you keep everything flowing downward out of the stomach instead of back up into the esophagus. Don't sleep propped up on a lot of pillows. This is just like sitting up in bed and increases intraabdominal pressure.

Try to eat frequent small meals rather than three large ones. Most people find they are most uncomfortable after dinner because this is usually the largest meal of the day.

Should I avoid particular foods?

Certain foods reduce the sphincter pressure at the lower end of the esophagus and favor reflux. These include chocolates, fatty foods, peppermint, and alcohol. Smoking also reduces sphincter pressure, causing reflux. Acidic fruit juices, such as tomato and grapefruit, may irritate the lining of the esophagus and stomach, while coffee, both regular and decaffeinated, stimulates stomach acid. These factors contribute to the symptoms of reflux esophagitis.

Are there any medications that worsen the problem?

Because the symptoms of reflux esophagitis are thought to be related to weakness in the sphincter muscle, any medications that relax this muscle would tend to aggravate the problem. Donnatal®, Librax®, and Bentyl®, drugs prescribed for irritable bowel syndrome (see Chapter 11), are just a few of the common medications that may have this effect.

Can emotions affect reflux esophagitis?
Absolutely. Tension, anxiety, anger, and stress will tend to increase the amount of stomach acid, thereby making it more likely that stomach acid will be regurgitated back up into the esophagus.

Are there any further considerations?
Yes. Even if you have reflux esophagitis, do not assume that every symptom in the area stems from these problems. If you have pain when you swallow food, or the food seems to get stuck while traveling down the esophagus, you may have another problem. It is important to bring these symptoms to your doctor's attention. If you suddenly develop pain in the area and it does not immediately respond to antacids, this also should be reported to your doctor. These are not typical symptoms of reflux esophagitis. Unfortunately, many people are not aware of this and will attribute any pain or discomfort in the chest or stomach area to reflux esophagitis. This can delay treatment of a more serious condition.

What is hiatus hernia?
There are two distinct kinds. More than 95 percent of people with hiatus hernia have a sliding (or direct) hernia. Under normal conditions, the stomach lies in the abdomen at the lower end of the esophagus. In a person with a sliding hernia, the upper portion of the stomach occasionally slides upward into the chest cavity through the esophageal hiatus (an opening in the diaphragm allowing the esophagus to pass through).

The other kind of hiatus hernia, known as paraesophageal hernia, is easily diagnosed on upper gastrointestinal (UGI) x-ray film examination. Unlike the sliding hiatus hernia, paraesophageal hernia is fixed in position and may require surgical correction even before symptoms occur. When the term hiatus hernia is used, it generally refers to the more common, sliding type.

What is the relationship between hiatus hernia and reflux esophagitis?

A sliding hiatus hernia was once thought to be the major cause of reflux esophagitis. After careful study and acid measurements in many people, it has been determined that the major factor contributing to reflux is the competence of the lower esophageal sphincter, *not* the presence of hiatus hernia.

9

APPENDICITIS AND MECKEL'S DIVERTICULUM

Anyone who has ever had abdominal pain has, at one time or another, asked himself or herself: "Could this be an attack of appendicitis?" Most people know that acute appendicitis, if not properly treated, can be a life-threatening emergency. When abdominal pain occurs, it is a matter of great concern.

Unfortunately, the symptoms of appendicitis are similar to those of many other problems, including ones outside the digestive tract. Even for a physician, the diagnosis is difficult. But you should be aware of certain signs so you will know when to bring a "simple bellyache" to your doctor's attention.

This chapter describes what appendicitis is, what its symptoms are, and what possible complications can occur. Another problem, Meckel's diverticulum, will also be addressed here; although it is far less common than appendicitis, it occasionally resembles appendicitis in location and pain pattern.

What is the appendix?

The vermiform appendix, which is its scientific name, is a small wormlike sac attached to the large intestine at the point where the large and small intestines are joined.

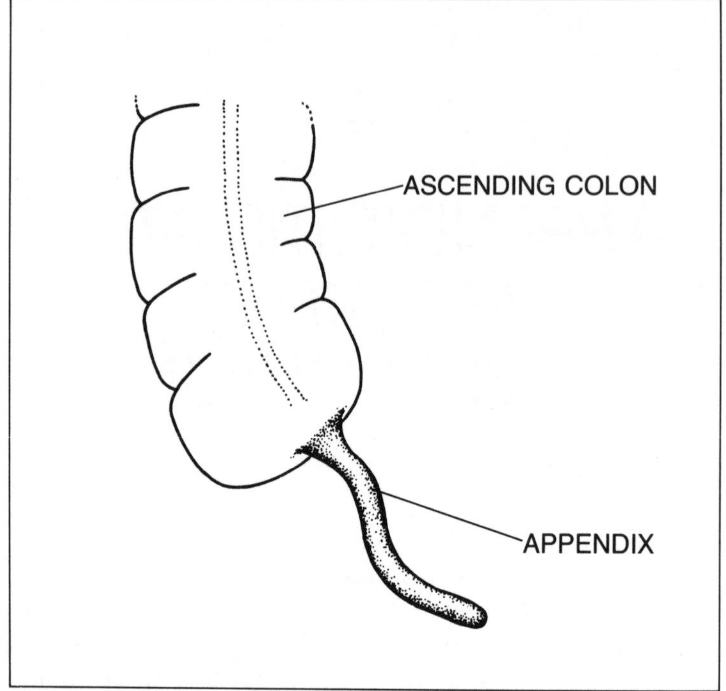

The appendix.

Does the appendix serve any purpose?

The appendix is richly endowed with lymph gland tissue that makes up part of the immune system of the digestive tract. However, because the digestive tract itself is so full of antibody-producing lymph cells, the appendix is not necessary for this function.

Why does it so frequently cause problems?

The answer is not known conclusively, but it's possible that its location and size make it vulnerable. Usually, a flare-up occurs when contents of the large intestine become lodged in the opening, causing compression of the blood vessels within the appendix that supply its cells with nutrients and oxygen. When this blood supply is cut off for a long enough time, the tissue cells that make up the appendix may die and eventually become gangrenous.

Appendicitis and Meckel's Diverticulum

What are the symptoms of acute appendicitis?
Appendicitis usually begins with a vague crampy pain around the navel. As the pain becomes more severe and constant, nausea and vomiting occur. A low-grade fever may be present, from 99°F to about 101°F. Within a few hours, the pain usually localizes to the lower right side of the abdomen and becomes increasingly severe. The pain is aggravated by coughing or sneezing—anything jarring the area. The abdominal muscles tighten in the affected area as the body attempts to protect itself, and the abdomen feels taut and rigid to the touch.

These are the textbook symptoms of appendicitis. If they occur in this sequence, the diagnosis is relatively easy to make. But often, because no two human beings are exactly the same, the symptoms vary enough to pose a diagnostic problem. Acute appendicitis may occur with all or only a few of these symptoms. The appendix may lie in a position that causes the pain to be felt in the back or on the left side, or it may remain in the umbilical area without ever localizing at all. In some individuals, especially older people, fever and pain may not be prominent; in some cases, these symptoms may be attributed to other conditions the person has, such as diverticulitis. This delays the diagnosis, resulting in an increased risk of complications.

For this reason, if you have any abdominal pain that lasts for even a few hours, call your doctor. He or she is in the best position to determine its cause.

Does appendicitis always cause nausea and vomiting?
Each person reacts differently to the same problem. Nausea and vomiting are common with appendicitis because any injury to the intestinal tract slows down its action. This prevents the normal processes of digestion from taking place. The results are loss of appetite, nausea, and vomiting as the undigested food is expelled rapidly from the body.

How does the doctor decide whether to perform surgery?

This can be a hard decision to make because diagnosis is often difficult. Many problems in the digestive system, as well as in other areas of the body, can produce symptoms similar to those of appendicitis. If the symptoms are not typical, a doctor has to determine whether the person really has appendicitis and needs surgery, or has another problem that could perhaps be treated with medicine. It is a crucial decision. If the doctor decides not to operate, and the person does have appendicitis, the appendix may rupture.

The surgeon must base the diagnosis on the information you provide, including the characteristic onset and distribution of abdominal pain, the delay in bowel movement, and the elevated temperature. In addition, the physician will rely on the blood count, which usually demonstrates an increased number of immature white blood cells. Interestingly, in older people, this elevation of white cells is often absent and, as previously mentioned, the pain distribution and fever are not as prominent. A plain abdominal x-ray is done that, in the presence of appendicitis, often demonstrates isolated stagnant air within the intestine. This x-ray film pattern is not specific for appendicitis, but indicates that there is an acute inflammatory process within the bowel that has reduced intestinal contractions. Referred to as an "illeus," this same x-ray film pattern occurs with other acute abdominal emergencies, such as diverticulitis, a perforated ulcer, or a gallbladder attack. The clinical symptoms, however, help to distinguish these conditions from appendicitis.

If there is still doubt as to the diagnosis, a barium enema is done. If the appendix does not fill with barium, the diagnosis of appendicitis is even more likely. In most cases, however, a barium enema x-ray film is not necessary.

Usually, if a surgeon feels the problem is most likely an appendicitis attack, he or she will go ahead and operate because the risks of not operating are far too great.

Why is it so dangerous for an appendix to rupture?
Initially, when the appendix becomes obstructed, infected, and inflamed, it is filled with bacteria and often fecal material. If it ruptures or bursts open, the bacteria and fecal material contaminate the entire abdominal cavity. This causes a serious infection called peritonitis—infection of the lining of the abdominal cavity—that is difficult to treat and can even be fatal.

Do people still die from appendicitis?
Yes. Even in our medically sophisticated age, approximately 2,500 people a year die from appendicitis. Many lives would be saved if people knew enough to seek medical attention in time.

Is an appendectomy a serious operation?
Any surgical procedure involves a certain degree of risk. But when performed by a skilled surgeon, an appendectomy is a relatively simple operation. The surgeon makes a small incision in the abdomen, carefully isolates and removes the appendix, making sure that none of the infected material spills out, and then sutures the wound. Most people are able to walk around the day after surgery and leave the hospital within a week.

Can appendicitis be treated medically?
Sometimes. If, for some reason, surgery cannot be performed immediately, the physician will administer intravenous antibiotics to control the infection until surgery can take place.

Is there any reason besides appendicitis to remove an appendix?
Surgeons will often routinely perform an appendectomy during an operative procedure for another problem in the area of the appendix. This is called an inci-

dental appendectomy; it is done because the appendix is so expendable, yet has the potential for causing serious problems.

Can someone have an appendicitis attack and not know it?

Yes. It's believed that many people have slight attacks of appendicitis that are never properly diagnosed. If the symptoms are not severe enough to warrant closer attention, such attacks may be passed off as gastroenteritis or an intestinal virus.

Is there such a thing as chronic appendicitis?

Yes. The medical name for this condition is recurrent appendicitis. What usually happens in such cases is that the appendix becomes inflamed and infected; the body, as a means of self-production, walls off the infection by isolating the appendix from the rest of the body through a complex process of scar formation.

Why is it dangerous to take a laxative when the appendix is inflamed?

It can be fatal to take a laxative when you have appendicitis. Any artificial stimulation of the digestive tract at this time could cause the appendix to rupture and, as we have seen, this is the event most to be avoided.

Is there any relationship between age and appendicitis?

Acute appendicitis is more common in adolescents and young adults than in patients over the age of fifty. However, these statistics should not delay the diagnosis of appendicitis in older persons. If symptoms of appendicitis are present, this diagnosis should be considered along with those of other diseases, more common to this age group, that produce similar symptoms.

What is a Meckel's diverticulum?

A Meckel's diverticulum is a congenital outpouching or pocket extending from the ileum. The ileum is the last

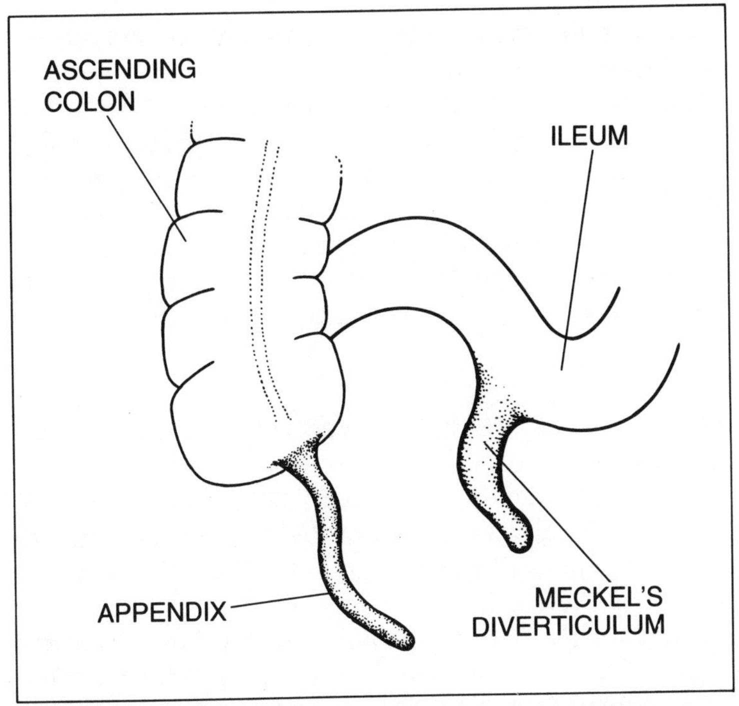

Position of Meckel's diverticulum in relation to appendix.

portion of the small intestine, and joins it to the large intestine or colon. Actually, a Meckel's diverticulum is a remnant of an embryonic duct that is normally present at a certain age in the fetus but recedes either before birth or a few weeks after birth. In most people it disappears completely, but in about 3 percent of the population it remains and can cause problems.

Why is a Meckel's diverticulum confused with appendicitis?

When a Meckel's diverticulum causes symptoms, it's usually in children or young adults. The symptoms of a Meckel's diverticulum are extremely uncommon in adults and particularly rare in older people. For this reason, when symptoms of a Meckel's diverticulum do occur in an adult, they may be attributed to more common conditions, such as appendicitis.

What are the symptoms of a Meckel's diverticulum?
A Meckel's diverticulum usually causes no symptoms at all. When symptoms do occur, the most common is rectal bleeding. The blood is bright red or maroon, and rarely black or tarry. Another, less common, symptom is pain in the right lower quadrant of the abdomen; this may be confused with the pain caused by appendicitis. Pain occurs when the diverticulum spontaneously twists upon itself or inverts into the intestine, causing blockage.

Why does a Meckel's diverticulum cause rectal bleeding?
For an unknown reason, about 20 percent of people with a Meckel's diverticulum develop ectopic mucosa within the diverticulum. This means that the inside lining of the diverticulum takes on the appearance of stomach cells. These cells may become functional and secrete hydrochloric acid, just as the cells in the stomach normally do. This accumulated acid may cause a peptic ulcer of the small intestine (ileum) adjacent to the diverticulum, resulting in pain and/or bleeding.

Is there any special test to locate a Meckel's diverticulum?
A Meckel's diverticulum is rarely seen on a conventional upper gastrointestinal series x-ray film. But another test, referred to as a nuclear scan, is often helpful. It takes advantage of the fact that stomach cells may be present in a Meckel's diverticulum. A radioactive dye called technetium is injected into an arm vein. This dye, which is quite safe, is unique in that it is selectively absorbed by stomach cells. Under normal conditions the dye should appear in the stomach, where it is absorbed, and in the bladder, where it is excreted from the kidneys. When the scanner machine shows the dye in the vicinity of the ileum, the assumption is that stomach cells are present in this atypical location. This confirms the diagnosis of a Meckel's diverticulum. The

scan, however, can be negative even when a Meckel's diverticulum is present.

How is a Meckel's diverticulum treated?

The only curative treatment for a symptomatic Meckel's diverticulum is surgical removal. When a Meckel's diverticulum is discovered incidentally during surgery for other conditions, it is usually removed, even if it has not caused symptoms.

10

DIVERTICULAR DISEASE OF THE COLON

Diverticulosis is rare in people less than thirty years of age but, by age sixty, 20 to 50 percent of people in this country develop it. It is the most common colon problem in people over forty years of age. Diverticular disease of the colon is characterized by many small outpouchings from the lining of the colon (large intestine); these herniated sacs are referred to as diverticula.

Diverticulosis is a silent disease, causing either no symptoms or symptoms so slight they're not even noticed. But once the disease is present, more serious problems may develop, including life-threatening complications.

Only since the turn of the century has diverticular disease become a problem in Western civilization. Before then, it was almost unheard of in medical literature. Diverticulosis is still unknown among people in primitive cultures. Because of this, researchers have come to some interesting conclusions about the causes of this disease.

What is diverticular disease of the colon?

Diverticular disease of the colon is frequently referred to as diverticulosis. The term is derived from the Latin word diverticulum, which means a small diversion from the normal path. This is exactly what characterizes the disease. After all the nutrients needed by the body are absorbed through the small intestine, the remaining substance moves down into the large intestine—a long hollow tube whose walls absorb excess water and electrolytes from the stool. Just prior to evacuation, muscular contractions move the stool along this tube into the rectum.

Diverticular disease usually occurs at the narrowest segment of the large intestine, called the sigmoid colon. Small segments of the intestinal wall herniate and form pouches, called diverticula, at the weakest points along the wall; these are the areas the nutrient arteries penetrate.

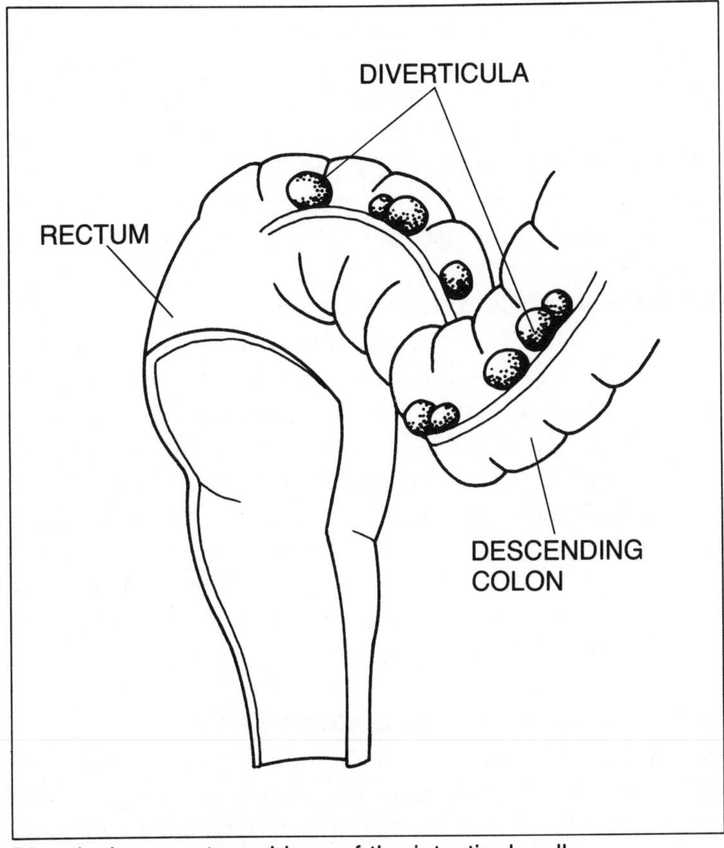

Diverticula, or outpouchings of the intestinal wall.

What is diverticulitis?

Diverticulitis is a complication occurring in fewer than 15 percent of people with diverticulosis. It refers to an infection and inflammation of one or more of the many diverticula or outpouchings of the large intestine.

What causes the diverticula to become infected and inflamed?

Until recently, it was thought that the infection and inflammation were due to small pieces of stool becoming lodged in one or more of the diverticula. However, stool often becomes lodged in the diverticula without causing any symptoms at all. It is now known that diverticulitis results from perforations of the diverticula; that is, an erosion occurs and the contents of the bowel escape into the surrounding tissue. When it is a minute perforation, called a microperforation, the infection and inflammation remain confined to a small area. In these instances, the intestinal contents are prevented from entering the abdominal cavity by the formation of a small abscess walling off the "leak." But if a large perforation occurs, called a macroperforation, there are far more serious consequences, including peritonitis.

Do the diverticula cause symptoms?

In some cases they do, but the symptoms are so mild they are not noticed, or they may be attributed to the irritable bowel syndrome (a condition, described in Chapter 11, in which the bowel becomes spastic). Diverticulosis is associated with constipation, diarrhea, or the presence of both those bowel disorders in an alternating pattern. There also may be a sense of bloating and distention, as well as cramping. Often, diverticular disease of the bowel is first diagnosed when an acute episode of diverticulitis occurs.

What are the symptoms of diverticulitis?

The main symptom of diverticulitis is pain, usually felt in the left lower portion of the abdomen. This is where the sigmoid colon is located and where the diverticula are most likely to appear. But the pain can occur on the lower right side or even in the center of the abdomen. The pain is usually accompanied by fever because of local infection. Nausea, vomiting, and chills may also occur during an attack.

Is diverticular disease hereditary?

No. The formation of outpouchings in the bowel that are indicative of the disease does not seem to be an inherited condition.

What causes diverticula to form?

The genesis of the disease is still not completely understood, but diet is believed to play a significant role in the formation of diverticula. A diet low in fiber causes harder stools, which move slowly through the intestinal tract. At the same time, gas is unable to pass smoothly down the intestinal tract and is trapped at different points along the way. It is this combination of events that generates excessive pressure inside the colon and causes the formation of the outpouchings, or diverticula. This is especially true in the sigmoid colon where isolated segmental areas of high pressure are the most common.

As mentioned earlier, diverticular disease was first described at the turn of the century. Interestingly, it was at about that time that refined food was introduced into the American diet, removing most of the fiber and bulk from our wheat products.

Is there any connection between diverticular disease and irritable bowel syndrome (IBS)?

Yes, in my opinion there is. It is often difficult to distinguish between these two conditions because the symptoms are so similar. Of course, there are people with diverticular disease who have never had any of the symptoms of IBS, just as there are people with IBS who never develop diverticula in the colon. However, these problems often coexist, with IBS present first, followed by the development of diverticulosis.

One of the important aspects of IBS is that it is not actually a disease but a functional disorder of the colon. In IBS, a disturbance exists in the normal peristaltic passage of gases and stool through the intestine. So far, evidence suggests that this disturbance may predispose some people to the formation of diverticula.

How is diverticular disease diagnosed?

Diverticula that do not cause any symptoms will be found only during an incidental diagnostic procedure. They are visible on an x-ray film taken by means of a barium enema, or can be seen through a colonoscope if they are present in the large intestine.

The diagnosis of diverticulitis can be difficult to make, especially if the pain appears in the right lower portion of the abdomen where the appendix lies. Occasionally, the correct diagnosis is made only after the surgeon has operated for what was thought to be appendicitis.

Is diverticular disease serious?

Diverticulosis, the presence of diverticula in the colon, is not in itself a serious condition, but it sometimes progresses to diverticulitis. If the diverticulitis is mild, the physician may recommend that it be treated at home with bed rest and antibiotics. More often, the doctor will hospitalize a person in order to allow intravenous feeding and give the bowel a rest. After a few days of rest and antibiotics, the inflammation and infection usually clear up. Sometimes, however, further complications set in.

What complications can occur with diverticulitis?

There are several possibilities. When a part of the digestive tract is injured, the entire intestinal tract can respond by slowing down or ceasing to work altogether. If bowel function ceases, the fecal material stagnates and blocks the intestine. This occurs in approximately 15 percent of all people with diverticulitis and is considered a surgical emergency.

When one or more diverticula become inflamed, a macroperforation may result, emptying fecal material directly into the abdominal cavity. Peritonitis may then set in, creating a serious emergency. Antibiotic therapy and surgical removal of the segment of colon containing the diverticula are often necessary when this happens.

Hemorrhage is one of the most dangerous complications of diverticular disease. The artery around an inflamed diverticula thins and finally bursts, resulting in severe loss of blood and vital fluids. If the bleeding does not stop on its own, emergency surgery must be performed. Another complication of diverticulitis is fistula formation.

What is a fistula?
A fistula is an abnormal communication between adjacent organs. When one or more diverticula of the colon become inflamed, the inflammation may spread outside the colon and adhere to any adjacent tissue. The most common type of fistula is between the colon and the urinary bladder. Called a vesicocolonic fistula, it is more common in men than in women. Intestinal contents may leak through the fistula tract into the urinary bladder. This causes symptoms of urinary infection, including urgency, frequency, and pain with urination. In addition, fecal liquid (fecaluria) and even air (pneumaturia) may appear in the urine with this type of fistula. Fistulas may also occur between adjacent loops and intestine, or they may extend from the colon into the vagina (colovaginal), or drain outside the body through the skin (colocutaneous).

Can diverticula be treated medically to prevent diverticulitis?
There is no drug available that can remove the diverticula once they have formed, but some things can be done to help control the symptoms of diverticular disease.

Some doctors recommend that their patients with diverticular disease avoid eating foods containing tiny pits or seeds; such particles may become lodged within the diverticula and cause an inflammation. This is a controversial point, however; as yet, there is no evidence to substantiate this advice.

The most important consideration in the treatment of diverticular disease is the addition of fiber to the diet. A high-fiber diet removes excess strain from the area by helping to move the stool rapidly through the colon, which is the basis of good bowel habits and the prevention of constipation.

11

IRRITABLE BOWEL SYNDROME

Of all the possible disorders of the digestive tract, irritable bowel syndrome, or IBS, for short, is the most prevalent—so prevalent that it's the greatest cause of industrial absenteeism in the country, greater even than the common cold.

IBS can be uncomfortable, worrisome, and, at times, disabling. But with all the unpleasant symptoms it produces, it causes neither tissue damage nor inflammation in any part of the intestinal tract. Because of this, IBS had been thought to be a psychological disorder—one way in which the body responds to emotional stress. Now researchers have established that it is a medical condition readily responsive to conservative medical treatment.

If you or someone close to you has IBS, this chapter will help you understand and deal with the condition. There are simple ways to minimize the symptoms and, in many cases, even eliminate them entirely.

Is irritable bowel syndrome ever called by another name?

Yes. Irritable bowel syndrome is really a technical term used to describe a certain set of symptoms. Its more common names include spastic colitis, mucous colitis, functional bowel disease, and irritable colon. Whatever it is called, the manifestations and treatment are the same.

What exactly is irritable bowel syndrome?

Irritable bowel syndrome is a functional disorder of the digestive tract. The symptoms arise as a result of a disturbance in the muscular propulsive action of the bowel, not because of inflammation or tissue damage. The bowel becomes spastic, with certain segments of it contracting too quickly and others too slowly. This can cause great distress and may account for most of the symptoms of IBS, including alternate periods of constipation and diarrhea. When the pattern of contraction speeds up, diarrhea results; when it slows down, the patient suffers from constipation. Other symptoms include gas pains, bloating, and distention. Gas becomes trapped within certain segments of the bowel because of the irregular propulsive pattern.

Sometimes the pain pattern of IBS is so severe it resembles more serious disorders. If the gas becomes trapped on the left side of the colon, the pain may radiate to the chest, left arm, or shoulder, resembling the pain pattern of angina. If the gas becomes trapped in such a way that the pain radiates to the back, it resembles the pain pattern of a kidney stone, gallbladder disease, or even appendicitis. Once the trapped gas is expelled, either by being passed or by means of a bowel movement, the pain usually subsides.

Do people with irritable bowel syndrome produce more gas than normal?

No. The amount of gas produced is about the same. What distinguishes people who have an irritable bowel from those who do not is the way the body handles the gas. This was demonstrated in a study in which an inert gas was infused into the intestinal tracts of a group of volunteer subjects (some had IBS and some did not). Although the amount of gas was the same, the people with IBS reported pain, distention, and bloating, while the other people had no symptoms at all and were able to expel the gas easily. It can be concluded from this study that people with IBS have a much lower tolerance for intestinal gas, and experience more discomfort from it, than does the rest of the population.

How is irritable bowel syndrome diagnosed?

Until recently, most doctors would encourage people with multiple bowel complaints typical of IBS to undergo a variety of tests to determine possible causes of the symptoms. Only when all the tests were negative would the diagnosis of IBS be made. This was called a diagnosis by exclusion.

Now we realize that it is not always necessary to subject a person to so many procedures in order to make the proper diagnosis. If a person has a long history of bowel problems typical of IBS, a diagnosis can be made on the basis of the history and a careful physical examination.

Of course, it is extremely important to assess any recent change in a person's normal bowel habits. Any change in character, frequency, consistency, or color of stool should be brought to the immediate attention of a physician. Also, the appearance of blood—either grossly present in the stool or on a trace examination in a doctor's office—would necessitate a thorough evaluation, including sigmoidoscopy (see p. 111), barium enema (see p. 112), and, possibly, colonoscopy (see p. 117).

When does irritable bowel syndrome begin?

IBS usually begins in early adolescence and will persist throughout life. A child may describe occasional episodes of constipation, diarrhea, or cramps, especially in times of stress. This is common, and does not imply the presence of IBS. However, when these symptoms persist into early adulthood, especially in the absence of stress, the diagnosis of IBS is likely.

How common is irritable bowel syndrome?
It is one of the most common minor medical problems. Almost everyone has had some degree of intestinal upset with emotional stress at one time or another. No matter how minor or transient these episodes were, they would still be labeled as part of IBS.

Is irritable bowel syndrome an emotional problem?
Most authorities now believe that IBS is a true physical disorder and not simply the result of psychological stress. Its symptoms can be aggravated by emotional stress, but there is an underlying physical defect in intestinal contractility and motility to account for the symptoms.

Might psychotherapy help in the treatment of irritable bowel syndrome?
If you suffer from the symptoms of IBS and feel they are aggravated by certain stress conditions in your life, by all means seek professional help. But the symptoms of IBS will probably not be cured and may not even be reduced by psychotherapy.

Is irritable bowel syndrome a serious condition?
No. Serious complications do not result from irritable bowel syndrome. It is only serious to the extent that it is temporarily disabling.

Can it be cured?
Probably not. If you have irritable bowel syndrome, you'll most likely have the problem all your life, though the symptoms will come and go irregularly. With our new understanding of IBS, physicians are now able to control many of the unpleasant symptoms, allowing the people who have this problem to lead a normal life.

Can it lead to cancer?
No link has ever been proved between irritable bowel syndrome and cancer of the colon or, for that matter,

any other life-threatening disorder of the intestinal tract.

Can it lead to any other serious disorder?

Some authorities suggest that people with IBS may be more likely to develop diverticular disease of the colon later on in life. This disease has been discussed at length in Chapter 10, but here is a brief description to show how the two problems may be related. Diverticular disease develops when small segments of the colon balloon out and form pouches. Most experts agree that these outpouchings, as they are called, result from isolated areas of high pressure within the lower part of the colon—pressure that can come from trapped gas and/or from stool moving through the colon too slowly. Trapped gas and constipation are symptoms of irritable bowel syndrome and, therefore, may lead to the beginnings of diverticular disease.

Are some people more likely than others to develop irritable bowel syndrome?

Yes. If someone in your family had IBS, you have an increased chance of developing it yourself. Irritable bowel syndrome is more prevalent in women than in men. Some researchers think it may be related to estrogen production in women because it often acts up before menstruation and during menopause, but this link is not definite. There also seems to be a connection between migraine headaches and IBS, especially in women.

Can milk and milk products cause irritable bowel syndrome?

No, but they may lead to symptoms that mimic IBS. Lactose intolerance often causes diarrhea, excessive gas, bloating, and distention, just as IBS does. For this reason, people who believe they're suffering from IBS should note whether their symptoms seem to occur after drinking milk or eating milk products.

Can diet foods cause irritable bowel syndrome?
No, they cannot create the actual condition, but many dietetic foods contain artificial sweeteners that can produce IBS symptoms. These sweeteners arrive in the colon in an undigested form and are acted upon by bacteria there, producing gas, bloating, and diarrhea. Once they are removed from the diet, the symptoms are completely eliminated.

What is the best treatment for irritable bowel syndrome?
Before I answer that question, I would like to mention a common method of treatment that doesn't work: the use of laxatives (other than fiber) and enemas. Many people with IBS have become addicted to these products, but their use can actually aggravate the condition and should be avoided.

Proper treatment of IBS involves exercise and diet. The most helpful exercises are those that strengthen the abdominal muscles, such as sit-ups. Stronger abdominal muscles are better able to handle the gas problem accompanying IBS.

Diet definitely plays a role in IBS treatment. The avoidance of fats seems to alleviate symptoms, although the reason for this is not completely understood. It is interesting to note that the same hormone (cholecystokinin) that is produced by the body in response to dietary fats may also stimulate contractions of the colon.

Probably the single most important factor in treating irritable bowel syndrome is the addition of fiber to the diet. Fiber helps return to normal the transit time of stool through the bowels and reduces intracolonic pressure. When constipation is the problem, fiber speeds up the transit time of the stool moving through the colon by producing a bulkier fecal mass. When the problem is diarrhea, the fiber slows down the transit time by facilitating absorption of water from the stool.

Onions, broccoli, cabbage, beans, and cauliflower are some high-fiber foods producing the greatest amount of gas. Keeping this in mind, experiment with these as well as other high-fiber foods to see which may bother you. Each person has individual sensitivities to different high-fiber foods.

Artificial fiber, in the form of bulk laxatives (see p. 36), often produces less gas than natural fiber. Therefore, if you have a problem tolerating high-fiber foods, you can supplement your food intake with the artificial fiber products on the market and gradually introduce natural fiber into your diet.

Remember, initially there may be a mild increase in gaseousness and distention with the introduction of any form of fiber into the diet. This is to be expected. Sometimes it can take several weeks for the body to adjust to an increased amount of fiber, so give yourself a chance. If your symptoms have not markedly improved after several weeks, try artificial fiber in place of natural fiber.

Patients often ask me how much fiber they should have in their diet. There is no simple answer because the right amount varies with each individual. Trial and error is the best way to find out how much and what kind of fiber will help you maintain a healthy colon without having such unpleasant side effects as gas, bloating, or distention.

Are any medications helpful in irritable bowel syndrome?

There are several prescription medications used to help relieve intestinal spasm and the discomfort associated with it. Some examples are Bentyl®, Librax®, and Donnatal®. However, the effectiveness of these medications in IBS is questionable, and long-term use can result in dependency. They are best reserved for isolated severe flare-ups, and then only under the supervision of a physician.

What is the fiber content of some common foods?

When patients are told to add fiber to their diets, they often think this means bran only. Actually, there are many different kinds of high-fiber foods beside bran, and they all help to make stool softer and bulkier and to speed up the transit time. This list should help you in choosing the right foods. Avoid products with added sugar, as these contribute to gaseousness as well as increasing caloric intake.

CEREALS	**DIETARY FIBER** (grams)
All Bran (1/3 cup)	8.8
Bran Buds (1/3 cup)	8.0
40% Bran Flakes (3/4 cup)	4.0
Corn Bran (2/3 cup)	5.4
Corn Flakes (1 1/2 cups)	3.5
Shredded Wheat (1 biscuit)	2.8
Wheat Chex (2/3 cup)	3.2
BREAD (1 slice)	
Pumpernickel	1.8
Rye (no seeds)	0.8
White	0.2
Whole wheat	1.4
VEGETABLES (raw except as noted)	
Beans	
Green, canned (1 cup)	4.0
Kidney, canned (1/2 cup)	17.9
Lima, cooked (1/2 cup)	8.3
Pinto, canned (1/2 cup)	13.0
Broccoli (1 cup)	7.0
Cabbage (1 cup)	4.2
Carrots (1 cup)	4.8
Cauliflower (1 cup)	3.2
Celery (1 cup)	2.1
Corn, boiled (1 medium ear)	3.1
Cucumber (1 medium)	1.5
Italian green peppers (1 cup)	1.0
Lettuce, iceberg (1 cup)	0.7
Peas, cooked (1 cup)	16.6
Potato, new, boiled (1 medium)	3.0
Spinach (1 cup)	0.2
Tomato (1 small)	1.0

Irritable Bowel Syndrome

	DIETARY FIBER (grams)
FRUIT	
Apple, delicious (1 small)	3.8
Apricots (2 medium)	1.8
Banana (1 small)	2.0
Cantaloupe (1/4 small)	0.9
Cherries (10 large)	1.1
Grapes (10 medium)	0.5
Orange, navel (1 small)	2.2
Pear, Bosc (1 medium)	4.8
Plums (3 small)	1.8
Raspberries (1 cup)	3.8
Strawberries (1 cup)	3.3
GRAINS (1/2 cup cooked)	
Barley grits	2.7
Brown rice	1.7
Buckwheat	1.8
Cracked wheat	4.1
Ground millet	1.1
Rolled oats	0.5

12

INFLAMMATORY BOWEL DISEASE

The term inflammatory bowel disease, IBD, for short, encompasses two intestinal disorders—ulcerative colitis and Crohn's disease. These diseases are among the most serious and least understood of all the problems of the digestive tract.

Over two million Americans suffer from the chronic and disabling effects of these diseases, effects that extend beyond the intestinal tract. Each year, 100,000 new cases are diagnosed in this country alone, and there may be more cases that have not been properly diagnosed.

Research into IBD continues as scientists look for answers regarding the cause and cure of these diseases. Medical practitioners, however, already have enough knowledge to enable most people with IBD to lead comfortable and productive lives. In the last twenty years, medicine and surgery have come a long way in treating, and even curing, some cases of these intestinal disorders. This chapter will familiarize you with the most common symptoms of inflammatory bowel disease as well as new types of treatment. As public awareness of IBD grows, more funds are allocated to the development of new drugs and innovative therapies to treat these diseases.

What is inflammatory bowel disease?

A wide variety of intestinal problems can result in an inflammation of the bowel, but, as noted previously, the term inflammatory bowel disease generally refers to two specific causes of inflammation of the bowel, ulcerative colitis and Crohn's disease.

Ulcerative colitis is an inflammatory disease limited to the large intestine, including the rectum. It does not involve the small intestine. Crohn's disease is also an inflammatory disease, but its bowel manifestations are more widespread. It creates small nodules or masses of inflamed tissue that penetrate deeply into the walls of the intestinal tract. This inflammatory process can occur anywhere from the mouth to the anus, although it often spares the rectum. Even when large segments of the intestine are involved, there are "skip" areas—patches of perfectly normal-appearing intestine interspersed with the areas of disease.

Crohn's disease seems to strike children and young adults more frequently than any other age group, but it can occur at any age. It was named after Dr. Burrill Crohn, the physician who, in 1932, first described the disease as an entity separate from tuberculosis of the intestinal tract. In 1981, I had the privilege of interviewing Dr. Crohn, who was then 98 years old. He described how difficult it was, initially, to differentiate Crohn's disease from other inflammatory bowel conditions.

This disorder is now referred to by different names, such as ileitis, granulomatous disease, regional enteritis, ileocolitis, and terminal ileitis. Often, the name used to describe Crohn's disease will depend upon the part of the intestine most affected. If only the ileum is affected, it may be called ileitis. If more of the small intestine is involved than just the ileum, it may be referred to as regional enteritis. If both the small intestine and the colon are involved, it may be called granulomatous ileocolitis. Often, Crohn's disease is referred to as granulomatous colitis or granulomatous disease because of the granulomas occurring in the intestine, which differentiate this disorder from ulcerative colitis.

What are the symptoms of ulcerative colitis?
The most frequent and obvious symptom of ulcerative colitis is diarrhea, with blood and pus in the stool.

Are the symptoms of ulcerative colitis the same in everyone?
No. The symptoms vary not only in their severity but also in their manifestations. For some people, ulcerative colitis comes on abruptly, with high fever, frequent loose, watery stool mixed with blood and pus, and strong abdominal cramps. For others, the onset is insidious, with mild abdominal discomfort and occasional bouts of diarrhea.

Five or 10 percent of people with ulcerative colitis have only one attack of the disease and no recurrences. In 65 to 75 percent of people with the disease, however, the pattern is a cyclical one of alternating flare-ups and remissions.

Ulcerative colitis can be divided into three categories: mild, moderate, and severe. A person with mild ulcerative colitis usually has no more than two to three bowel movements a day, mild cramps, and intermittent bleeding, if bleeding is present at all. Only a small portion of the colon is involved, and there are rarely any other manifestations of the disease outside the intestines.

About 25 percent of people with ulcerative colitis have a moderate form of the disease. Here, the symptoms are a little more disabling but do not present any major problems.

Severe ulcerative colitis is a serious disease. There are often more than six bowel movements a day and enough bleeding to cause anemia. There is high fever, loss of appetite, loss of weight, and severe cramps. In such severe cases, there is always the possibility of further complications.

What happens to the colon to produce these symptoms?

In ulcerative colitis, the lining of the colon becomes inflamed and gradually breaks down into multiple, raw, weeping ulcers. The ulcers erode into the underlying veins and capillaries and begin to bleed. The inflamed intestinal lining also exudes the mucus and protein appearing in the stool. As the body attempts to heal the ulcers, scar tissue forms, which can shorten and narrow the colon, thus impairing its ability to properly absorb water from the fecal material.

The inability of the inflamed colon to absorb water from the stool and the disturbance of contractions are all symptoms of ulcerative colitis. Also, the resulting inflammation of the rectum causes a frequent urge to defecate.

At what age is a person likely to get ulcerative colitis?

The incidence of ulcerative colitis seems to peak twice, once in the teen years and then again in the fifth decade of life. Few cases are found in people over the age of sixty-five.

Can anyone develop ulcerative colitis?

Yes. Statistics show that no ethnic or socioeconomic group is more or less likely to develop the disease than any other group. You do have a greater chance of having the problem if someone in your family has inflammatory bowel disease.

How do I know if I have ulcerative colitis?

If you have diarrhea lasting more than seventy-two hours or see blood in your stool, consult your doctor. If your physician suspects you might have ulcerative colitis, he or she will recommend some diagnostic tests. The initial test is often a sigmoidoscopy. This is a simple procedure, done in the office, where the doctor inserts a small, illuminated tube into the rectum. Through the tube, your doctor can see into the lower part of the colon and the rectum. If necessary, a tissue sample can be taken for a biopsy. If you do have ulcerative colitis, the lining of the rectum and colon will appear inflamed and possibly ulcerated. The biopsy will confirm the presence of IBD.

Your doctor may also request that you have a barium-enema x-ray film to help define the extent of the disease. Usually, this is done in a radiologist's office. A radiopaque dye is introduced into the rectum through a short enema tube; the dye passes through the entire large intestine, making its features—and the extent of the disease—visible on an x-ray negative.

Are there any diseases that seem like ulcerative colitis but really are not?

Yes. In fact, it is now believed that certain cases of ulcerative colitis of a few years ago may really have been false diagnoses. This is especially true in instances where there was only one episode of the disease.

Campylobacter is a bacterium that may cause bloody diarrhea and high fever, symptoms almost identical to those of ulcerative colitis. The effects of *Campylobacter* are generally not chronic and they respond readily to antibiotic therapy. There are other intestinal infections, as well as certain antibiotic-related diarrheas, which resemble ulcerative colitis but are not of chronic nature and can be successfully treated.

Arteriosclerosis develops with increasing age and may affect the blood vessels of any organ within the body. When arteriosclerosis affects the blood vessels in the colon, a type of bloody diarrhea may occur as a result of inadequate blood supply. Known as ischemic colitis, this condition is often mistaken for ulcerative colitis.

What problems does ulcerative colitis cause outside the intestinal area?

Frequently, people with bowel problems also have problems in other areas of their bodies. Skin lesions may exist; rheumatoid arthritis as well as ankylosing spondylitis (a form of spinal arthritis) may also occur, as may uveitis and keratitis, two serious eye conditions.

Some people with ulcerative colitis also have a lactose intolerance due to a lactase deficiency. Without lactase, milk and milk products cannot be digested. In these cases, diet must be adjusted accordingly.

What are the possible intestinal complications of ulcerative colitis?

The intestinal complications may be mild, ranging from bloody diarrhea to localized abscesses and strictures (abnormal narrowing of the intestine). But there can also be more serious complications. If one or more of the colonic ulcers erodes into a major artery, hemorrhage may result, just as in peptic ulcer disease. Complete perforation through the wall of the large intestine can also occur, releasing bacteria and fecal material into the abdominal cavity. Should this happen, peritonitis results, which can be a life-threatening problem. If enough scar tissue forms, obstruction of the lower bowel will block the passage of waste material out of the colon and prevent the normal continuation of the digestive process. Any of these conditions warrants surgery.

Does ulcerative colitis cause cancer?

People who have had extensive ulcerative colitis (that is, colitis involving the entire colon) for ten years or more are at an increased risk of developing colon cancer. In fact, the risk of colon cancer increases by 20 percent for every ten years a person has this condition. To make matters worse, the type of malignancy that develops—which is within the flat portion of the intestinal wall—is harder to identify visually than bowel cancer—which develops in the form of a polyp, or protrusion from the intestinal wall (see Chapter 13).

Because ulcerative colitis causes a wide array of bowel symptoms, it is often difficult to diagnose an early cancer; yet, the development of cancer is a real risk. Therefore, surgical removal of the colon is sometimes advised in people who have had symptomatic disease of the entire colon that has persisted for ten years or more.

How is the surgery done?

The operation for ulcerative colitis is called a colectomy. The surgeon cuts into the abdominal wall at the point where the ileum (the end portion of the small intestine) lies, and then brings the cut end of the ileum outside the body through a small hole in the abdominal wall. The end of the ileum is sewn to the skin, and a new passageway, called an ileostomy, is created to act as a reservoir to collect fecal liquid wastes. At the same time, the entire colon is removed, thereby curing the person of the disease (see Chapter 18).

Is there any alternative to this procedure for patients with ulcerative colitis?

A surgical procedure has been devised recently that preserves the muscles and lower portion of the rectum. First the entire diseased colon is removed, excluding the lower rectum. The diseased surface lining of the rectum is then taken out, leaving the remaining rectal tissue and muscles intact. The next step is to attach the ileum to the remainder of the lower rectum with the creation of a pouch (similar to an ileostomy) located just above the anus. The advantage of this operation is that it maintains the normal process of anal defecation by preserving bowel continuity. Experience with this operation is currently limited; long-range observation for complications is now underway at several medical institutions.

How is ulcerative colitis treated when surgery is not necessary?

The answer depends on the severity of the disease, the frequency of attacks, and the person's overall health. Diet, once considered an important tool in treatment, is no longer thought to play that significant a role. But good nutrition is important because the disease may cause a deficiency of certain important nutrients.

Except for the need to exclude milk and milk products for those people with lactose intolerance, there is often no reason to recommend a restricted diet, although a low-fiber diet is often advised when the colitis is active. If a person finds that one particular food causes symptoms to intensify, that food or food group should be avoided. Generally, most foods are well tolerated.

Steroids have proved helpful in controlling flare-ups. In cases where the lower part of the rectum is affected, steroid enemas can be used to treat the inflammation. Sulfasalazine (Azulfidine®) in tablet form is used as long-term maintenance therapy during remissions to prevent future attacks. Antidiarrheal medications, which were once used to help control symptoms, are best avoided because they can mask some early symptoms of the disease.

Are the symptoms of Crohn's disease similar to those of ulcerative colitis?

Often, especially in the early stages of Crohn's disease, the symptoms may be very much alike, including abdominal cramps, bloody diarrhea, fever, and weight loss. But as the disease progresses, certain signs and symptoms occur to distinguish it from ulcerative colitis.

Only 15 percent of people with Crohn's disease have blood in their stool, whereas almost all people with ulcerative colitis do. Also, more than half the people with Crohn's disease initially have pain and feel a lump in the lower right side of the abdomen. Sometimes the symptoms are so similar to those of appendicitis that an operation is performed. Only then is the diagnosis of Crohn's disease made. One of the common problems in Crohn's disease is fistula formation that can occur anywhere in the intestinal tract. Fistulas are rare in ulcerative colitis and occur, if at all, in the rectovaginal area in women with the disease.

Are the symptoms of Crohn's disease the same in everyone?
No. The symptoms depend on the extent of the disease, the area of the intestinal tract that is affected, and the overall health of the person.

What happens in the intestine to cause the symptoms of Crohn's disease?
The areas of the intestine affected by the disease develop multiple small nodules, with a granular appearance, accompanied by multiple linear ulcers that penetrate deep into the walls of the intestine. The inflammation and the impairment of bowel function that accompanies it produce watery diarrhea and abdominal cramps.

What are fistulas?
Fistulas are abnormal tracts or passageways leading from one part of the inflamed intestine to another part, to a nearby organ, or to the external skin surface.

When the outer, thin, watery lining of the intestine becomes inflamed, abscessed, and covered with exudate, the intestinal wall festers and adheres to the adjacent bowel or to another close structure, such as the bladder. When this happens, a tract fistula forms, and carries infected contents to the nearby affected organ. (The development of fistulas often confirms the diagnosis of Crohn's disease.)

How can I tell if I have Crohn's disease?
If you have diarrhea or cramps lasting for more than seventy-two hours, it's best to consult your physician. If your doctor suspects, on the basis of a physical examination and a careful history, that you might have Crohn's disease, diagnostic tests will be performed.

The tests will depend on the location of symptoms. Initially, because Crohn's disease usually attacks the small intestine first, your doctor may suggest an upper gastrointestinal series x-ray film, or UGI. For this simple test, you will be asked to drink a radiopaque solution so that your upper intestinal tract will be visible on a series of x-ray films. Your doctor may also order a barium enema to determine the existence or extent of the disease in your lower bowel.

In order to confirm the diagnosis, your physician may want to take a small biopsy from your intestinal tract. He or she may perform a sigmoidoscopy (an inspection of the walls of the colon through a tube) or, if the disease is higher up in your bowel, a colonoscopy, which is a relatively simple and painless test. Your doctor will give you mild sedation, then introduce a long, flexible fiberoptic tube into the rectum. This tube is illuminated at one end and enables the doctor to see into the large intestine throughout its entire length. A biopsy of the colonic tissue and photographs of any areas of inflammation can be taken.

What complications can occur with Crohn's disease?
The four most serious complications of Crohn's disease are the same as those of ulcerative colitis: perforation, obstruction, hemorrhage, and intractability (the failure of the disease to respond to medical therapy).

Additional complications particularly characteristic of Crohn's disease are the fistulas and abscesses, or inflammatory masses, that occur when a large segment of intestine becomes inflamed. Many people with Crohn's disease of the ileum develop kidney stones, gallstones, or arthritis, including ankylosing spondylitis (a form of spinal arthritis). Eventually, many people with Crohn's disease develop problems of the anus, including fissures and abscesses.

Because Crohn's disease usually involves the small intestine, an organ responsible for the absorption of nutrients, malnutrition may also develop.

Is there as great a chance of developing bowel cancer with Crohn's disease as with ulcerative colitis?
An individual with Crohn's disease runs a slightly greater risk of developing bowel cancer than does a

normal individual. But just how much more of a chance is still uncertain. Most researchers believe that, although the incidence of colon cancer is increased in patients with Crohn's disease, it is not as high as with long-standing ulcerative colitis.

How is Crohn's disease treated?

Depending on the severity and the extent of the disease, Crohn's disease is treated in much the same way as is ulcerative colitis. Steroid therapy is still the treatment of choice during the flare-ups. Sulfasalazine (Azulfidine®), the antibiotic that seems to help prevent attacks of ulcerative colitis, is used in much the same way in the treatment of Crohn's disease, but is not as effective. In the last few years, metronidazole (Flagyl®) has been used in high doses in people, both intravenously in hospitals and orally outside the hospital setting, to treat the perianal fistulas frequently accompanying Crohn's disease (those are the fistulas surrounding the anus). So far, the evidence is inconclusive as to how effective this drug really is.

When is surgery necessary for Crohn's disease?

Surgery for this disorder is different than for ulcerative colitis. Once a colectomy (the total removal of the colon) is performed on people with ulcerative colitis, they are cured. In contrast, there is no surgical cure for Crohn's disease since it may involve not only the large intestine but also the small intestine. Whereas the colon may be removed entirely without any ill effects, the small intestine is necessary for life. Once a segment of the small intestine with Crohn's disease is removed surgically, the disease may recur in the remaining portions of the small intestine at any time. In fact, surgery for Crohn's disease may hasten this recurrence. Continued surgical shortening of the small intestine will result in severe nutritional problems. Therefore, surgery is done only to deal with a problem that cannot be handled with ordinary medical treatment.

Inflammatory Bowel Disease

How common is inflammatory bowel disease?

It is estimated that in the United States one out of every thousand adults has ulcerative colitis and one out of every three thousand has Crohn's disease. From all available information, we can see that inflammatory bowel disease is on the increase, with Crohn's disease increasing even more rapidly than ulcerative colitis.

What causes inflammatory bowel disease?

No one knows for sure, but researchers are studying several different possibilities. Inflammatory bowel disease may be caused by an infectious agent, such as a virus or bacterium present in the environment but not yet identified. There is also a possibility that people with inflammatory bowel disease have a defect in their immunologic systems that prevents them from fighting off the particular infectious agent causing the disease. Some experts think inflammatory bowel disease may be the result of an "autoallergy." This means that IBD patients are producing antibodies that attack and damage their own tissue, especially in the bowel. So far, no one has arrived at a conclusive answer.

Since the cause of inflammatory bowel disease may be infectious, is there any possibility the disease is contagious?

Absolutely not. Many studies have been done on people with IBD to see if organisms are transmitted from one person to another. All the studies have come to the same conclusion: such transmission does not occur. Although the disease does appear in families, it only appears in family members who are related by blood. Spouses, adopted children, and adopted siblings are no more likely to get IBD than are other people in the population.

Is inflammatory bowel disease inherited?

Yes and no. There is no way to predict with absolute certainty that a child will have IBD, even if both parents have it. Although it does run in families, sometimes it will skip generations; in some families, each generation has at least one member with the disease. Inflammatory bowel disease does not consistently follow any known heredity pattern. If you have inflammatory bowel disease, the chances that some other member of your family has it are five times greater than if you didn't have it.

Can emotional factors cause inflammatory bowel disease?

There is absolutely no evidence to indicate that emotional factors can cause inflammatory bowel disease. Several years ago, doctors at Johns Hopkins University studied a group of people with inflammatory bowel disease to find out whether any specific personality type was more likely to develop these disorders. They compared the personality characteristics of people with IBD against healthy subjects. They concluded that both groups demonstrated the same wide array of personality characteristics, and that no specific personality profile could be drawn of a person with inflammatory bowel disease.

But it has been shown that when inflammatory bowel disease is present, emotional factors can aggravate the disease as well as precipitate an attack. The disease itself can be the cause of emotional problems. Many people have difficulty dealing with the sometimes disabling and often embarrassing ramifications of these disorders. And, of course, any chronic illness brings with it a certain amount of psychological and interpersonal, and even financial, stress.

Would psychotherapy be recommended in treating the disease?

Inflammatory bowel disease is often a recurring problem; it creates stress and is affected by stress. Psychotherapy may help a patient learn how to cope better with the problems associated with IBD.

What about chemotherapy to treat inflammatory bowel disease?

Two cancer chemotherapy drugs, azathioprine (Imuran®) and 6-mercaptopurine (Purinethol®) are being used in the treatment of inflammatory bowel disease. Although these drugs can cause serious side effects, they seem to be relatively safe when taken in small dosages and effective in reducing symptoms.

Where can I get further information about inflammatory bowel disease?

The Resource List in this book lists organizations that make information available to the public on all conditions and diseases of the digestive tract. If you or someone close to you has inflammatory bowel disease, you may want to contact The National Foundation for Ileitis and Colitis to discuss any medical, psychological, or financial problems you may be having. If they cannot help you directly, they will refer you to an agency or individual who can. They have many booklets about the disease as well as a monthly newsletter reporting new findings in the field.

13

INTESTINAL POLYPS

Polyps are abnormal growths of tissue that can occur almost anywhere in the body, including the intestinal tract. Because intestinal polyps have a tendency to undergo malignant changes at some point in their growth, their early detection and removal can make a big difference in overall health and survival. Unfortunately, intestinal polyps—which most often occur in the colon—usually don't cause any symptoms until they have grown to dangerous proportions. But there are simple ways to discover their presence long before symptoms occur. Once they are detected, they often can be removed without surgery. Understanding polyps and knowing what signs to look for are the best defenses against cancer of the colon.

Are all intestinal polyps the same?

No. There are many types of intestinal polyps; they are classified on the basis of their appearance, growth pattern, and malignant potential. The two main categories of polyps are pedunculated polyps, which look just like a mushroom or cherry on a stalk, and sessile polyps, which lie flat against the intestinal wall and do not have a stalk.

Intestinal Polyps

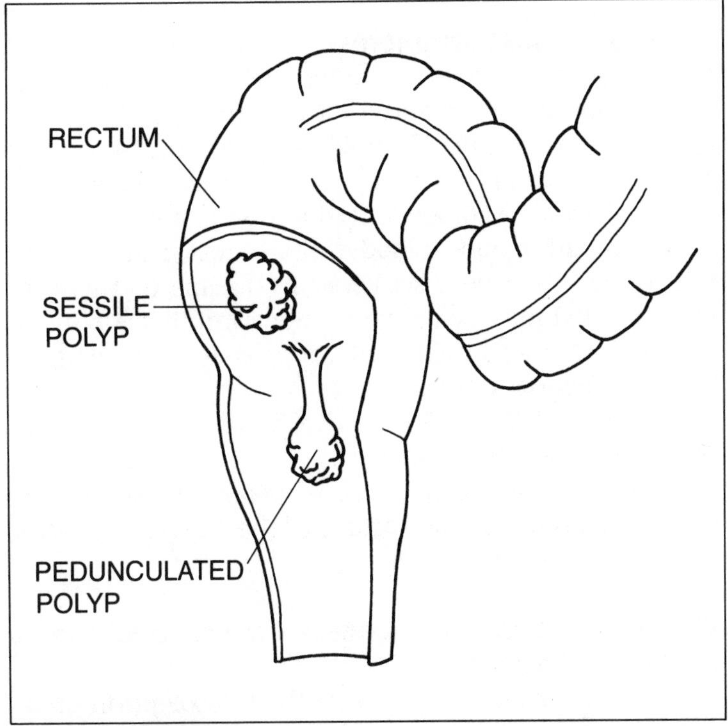

Intestinal polyps.

Although every polyp has the potential to become malignant, larger sessile polyps are more frequently found to be malignant than smaller, pedunculated ones. Malignant sessile polyps continue to spread by growing on the intestinal wall.

What causes polyps?

It is not known why polyps form in the large intestine, just as it cannot be fully explained why smaller growths occur elsewhere in the body. Certain kinds of polyps are hereditary. Two examples of hereditary colon polyposis are familial polyposis and Gardner's syndrome. With these disorders, hundreds or thousands of intestinal polyps develop and symptoms appear at an early age. Most people with hereditary colon polyposis have colon surgery before the age of thirty.

Do polyps cause symptoms?

Generally, nonhereditary polyps do not cause symptoms. They are discovered either during a diagnostic evaluation for another problem or during a routine physical examination when the physician chemically tests the stool for blood. Often a minute amount of so-called occult blood—blood that is not visible to the naked eye—will be discovered by means of this test. Occult blood may be an early indication that a polyp is present. This sign will lead the physician to suggest further diagnostic tests.

Why does a polyp bleed?

The blood found in the stool is usually caused by the sloughing off of the top surface of the polyp as the stool moves through the colon.

What is the relationship between polyps and cancer in the intestinal tract?

It is generally believed that, with the exception of ulcerative colitis, almost all colon cancers (that is, carcinomas) begin as polyps. If a polyp remains undiagnosed, it will continue to grow slowly. Once it reaches more than 1 cm in size, it has an increased chance of becoming malignant. As mentioned before, some polyps have greater potential for becoming malignant than others; some polyps, if they do become malignant, will spread the disease faster than others.

How are polyps treated?

Most polyps, once discovered, should be removed. The method of removal depends on the type of polyp. A pedunculated polyp, the kind that sits on top of a stalk, can often be removed through a colonoscope. A colonoscope is a long, flexible tube made up of many fiberoptic bundles that transmit light. The tube has a small diameter and is easily passed from the rectum into the large intestine. It allows the physician to actually see the inside of the intestine and to direct the inside tip with controls at the opposite end. A channel through the colonoscope allows the passage of biopsy forceps to obtain specimens and of wire loops to cauterize polyps and remove them. After this procedure, the stalk and polyp are carefully examined microscopically by a pathologist to see if there is any malignant tissue. If there has been an infiltration of malignant cells into the stalk, then limited surgery may be required. But this is rare. If a larger sessile polyp is found, then surgery is generally necessary to remove it. The extent of the surgery depends upon whether or not malignant cells are present and, if so, how far the malignancy has spread.

Can a polyp grow back after it has been surgically removed?
No. Once the polyp has been completely removed, either through a colonoscope or by surgery, it does not grow back. But polyps may develop elsewhere in the colon, so continued medical surveillance is necessary.

Is there anything special I should do if I have a polyp?
Yes, there is. Polyps are a recurring disease. Once you have had one polyp in your intestinal tract there is always a possibility of developing others. If you're in this category, you should have, at regular intervals, a stool test for occult blood, a barium enema, and colonoscopy.

Are intestinal polyps more common in older people?
Polyps of the large intestine increase in incidence with age. They are also more common in individuals with a family history of colon polyps or cancer of the colon.

If I have external polyps, such as nasal polyps, am I more likely than most people to develop polyps in the intestinal tract?
No. There is no known relationship between any other kind of polyp and the ones that proliferate in the intestinal tract.

14

PANCREATITIS

One out of every thousand hospital admissions is for the treatment of acute pancreatitis, an inflammatory disease of the pancreas. Despite this statistic, few people have ever heard of this disorder.

Pancreatitis can lead to mild or severe illness, with victims of milder attacks recovering in a few days and victims of severe attacks dying in a matter of hours. Unfortunately, there is little doctors can do to cure pancreatitis or even to treat it. But much can be done to prevent this disease.

What is the function of the pancreas?

The pancreas is probably the most powerful digestive organ in the body. Because of this, any significant damage to it can be serious. Not only are the pancreatic enzymes vitally important for the digestion of fats, carbohydrates, and proteins, but the pancreas is responsible for secreting insulin, which is necessary for the proper utilization of glucose; diabetes mellitus results when insulin is absent or insufficient.

Nevertheless, a person can survive without a pancreas, and surgical removal of an extremely damaged pancreas is sometimes unavoidable. After such surgery,

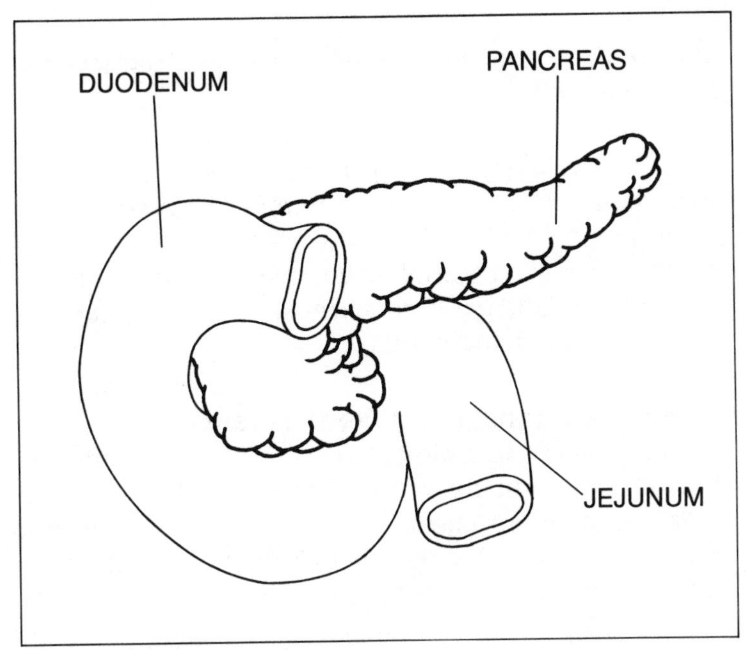

The pancreas.

a person develops diabetes mellitus and requires large amounts of pancreatic enzymes with each meal to replace the missing natural enzymes. General nutrition is impaired because artifical enzymes are not as efficient as natural ones in breaking down foodstuffs into absorbable forms.

What is pancreatitis?
Pancreatitis is an inflammation of the pancreas gland and, in most instances, of the pancreatic ducts as well.

What causes the pancreas to become inflamed?
Sometimes a gallstone becomes lodged in the pancreatic ducts, preventing the free flow of pancreatic enzymes into the duodenum. These enzymes are then absorbed by the surrounding unprotected pancreatic tissue and autolysis sets in—that is, the gland digests itself. This is a medical emergency. In rare instances, pancreatitis can be caused by a malignant growth in the pancreas. It may also be caused by the intake of

alcohol or certain medications, including tetracycline, and some types of diuretics.

Who is most likely to get pancreatitis?
Pancreatitis is more common in men than women and usually strikes after the mid-thirties. It is most prevalent in people who drink excessive amounts of alcohol over a long period of time; because of this, it is referred to as alcoholic pancreatitis.

How does alcohol affect the pancreas?
Alcohol may have a direct toxic effect on the pancreas, but only after many years of drinking does the pancreas become chronically inflamed. Alcohol may also damage the pancreas by other mechanisms not clearly understood.

Is pancreatitis a serious disease?
Pancreatitis can be a serious, even fatal, disease. Usually, the first attack responds well to aggressive medical treatment; more than 75 percent of subsequent attacks resolve within three or four days after the initial attack.

What are the symptoms of pancreatitis?
The major symptom is usually quite dramatic: severe pain around the navel, penetrating straight through to the back. The pain may be slightly relieved if you bend forward at the waist, and will intensify if you lie flat. This is an important indicator that helps the physician make the diagnosis of pancreatitis. There may also be nausea, vomiting, fever, chills, and profuse sweating.

How is pancreatitis diagnosed?
The pain pattern of pancreatitis is so unique that a physician can usually make the diagnosis after a brief examination of the patient. If the physician strongly suspects pancreatitis, he or she will likely hospitalize the patient to ensure proper treatment. Blood and urine

tests will then be done to confirm or rule out the diagnosis. (When the pancreas is inflamed, pancreatic enzymes escape into the bloodstream and urine.)

Even after the diagnosis is firmly established, however, the physician must look further for the cause of the acute flare-up. The next step is usually to investigate the gallbladder and the bile duct to see if an obstruction by a gallstone is the cause of pancreatitis.

What are the complications associated with pancreatitis?

One result of long-standing pancreatitis is an insufficient production of pancreatic digestive enzymes, leading to incomplete digestion and a consequent shortfall of nutrients. This situation is characterized by severe diarrhea, an indication that the pancreas is no longer capable of properly digesting dietary fats and other foodstuffs.

Hemorrhagic pancreatitis is a serious complication that can prove fatal. The blood vessels in and around the pancreas erode, causing profuse bleeding within the gland as well as in the surrounding tissues. Another complication is a pancreatic pseudocyst. This is a localized collection of fluid within the pancreas that may become infected or form an abscess. Any one of these complications can become a life-threatening emergency.

How is pancreatitis treated?

In an acute attack of pancreatitis, treatment is directed toward resting the pancreas by reducing the need for pancreatic secretions as much as possible. The person is hospitalized and fed fluid nutrients intravenously. It is sometimes necessary to insert a tube through the nose into the stomach to drain the gastric secretions and decompress the stomach. Some physicians administer antibiotics to prevent the development of an infection or treat one already present. Strong painkillers are also given to the person until the attack subsides.

How long does it take an acute attack to pass?
Usually, if gallstones are not obstructing the pancreatic ducts, and the person is being properly treated, an acute attack will subside within three or four days.

After an acute attack, is it possible for the pancreas to return to normal?
If gallstones are the cause of pancreatitis, the pancreas will return to normal once the stones have been removed and the inflammation subsides. But in pancreatitis brought on by excessive consumption of alcohol (which is far more common), the pancreas may remain inflamed even after the attack, although the acute symptoms may disappear.

Does this mean a person with alcoholic pancreatitis can get another attack?
Yes. If someone with alcoholic pancreatitis continues to drink, the symptoms will most likely recur.

If I've had an attack of acute pancreatitis, should I change my life-style in any way?
Once the inflammation has subsided, you can return to a normal diet unless some degree of pancreatic insufficiency has resulted from the attack. If drinking is the cause of the pancreatitis, you must stay away from alcohol. If pancreatitis has been caused by a medication, that medication must be avoided.

15

INTESTINAL INFECTIONS

Nowadays, when a great deal of attention is focused on heart disease and cancer, people easily lose sight of the fact that a majority of our medical ills are still caused by infectious organisms. Viruses, bacteria, and protozoa account for a variety of symptoms and diseases, ranging from the common cold to pneumonia. They can attack any organ in the body and, depending on the health of the individual, can create problems running the gamut from relatively minor symptoms to life-threatening illnesses.

All too often it is the intestinal tract that becomes host to these trespassers; they enter through the mouth and lodge themselves in the body. As the body attempts to fight them off, we are made miserable by the unpleasant symptoms of nausea, vomiting, cramps, and diarrhea. Sometimes, no matter how stubbornly the system attempts to ward off infection, the invaders prevail and illness ensues. Fortunately, when natural immune mechanisms are not strong enough to maintain health, a backup army of antibiotics and other drugs can provide reinforcement.

Infectious gastroenteritis is the catchall term for invasions of the intestines by foreign disease-causing organisms, whether the resulting infection is a mild case of food poisoning or a severe case of dysentery. The initial symptoms are often similar, but the overall problems resulting from these infections are not.

This chapter describes the kinds of infectious organisms that cause gastroenteritis, and the symptoms and course of illness they produce. A little knowledge goes a long way toward preventing many intestinal infections and toward managing more serious problems if infection does take hold.

What are the symptoms of mild infectious gastroenteritis?

The symptoms are a slight intestinal upset accompanied by cramps, diarrhea, loss of appetite, nausea, and vomiting. There may also be a low-grade fever of about 101°F. All those symptoms are not present in every person, but most of them usually are. The most important point is that with mild gastroenteritis, symptoms totally disappear within a day or two.

What should I do if I develop these symptoms?

Even mild infectious gastroenteritis will probably make you feel quite ill. Take it easy, and stay in bed for a day or two until the symptoms pass. If you have diarrhea or vomiting, keep in mind that dehydration—the loss of body fluid without adequate replacement—can result. It is important that you continually replace the fluids you lose and take in some sugar with the fluids. (Sugar facilitates the intestine's absorption of water and salt; juice or sweetened tea does the job quite well.) In older people, and in debilitated people, dehydration can be particularly dangerous. Within hours of a sudden loss of fluids from vomiting or diarrhea, dizziness, a drop in blood pressure, fainting, and even death can occur. If it's a warm summer day and you've been out in the sun, water loss—in addition to vomiting or diarrhea—is more rapid than on a cool day, so even more fluids must be replaced.

If you are taking a diuretic medication (one that

tends to increase the flow of water out of the body), dehydration can be especially serious. People on such medications should contact their doctor immediately if they develop symptoms of gastroenteritis.

What should I eat during a bout of mild gastroenteritis?

Usually, mild infectious gastroenteritis will take away most of your appetite. Don't try to force yourself to eat. You can easily survive for a few days without any food at all, as long as your fluid intake is adequate. If you feel nauseated or are vomiting, the best thing to do is avoid eating. As you begin to feel better, your appetite will return, and you can start to eat. Keep your diet light; eat only when you are hungry; and drink fluids.

Are any foods to be avoided?

A person who suffers even a mild attack of infectious gastroenteritis may become temporarily unable to properly digest milk and milk products, such as ice cream or cheese. As a result, these foods may create large amounts of gas and perhaps even worsen the diarrhea.

Are any over-the-counter medications useful for stopping diarrhea?

No such medications treat the actual infection. All they do is slow down diarrhea, and this is not always beneficial. The more you slow down diarrhea, the longer the infectious agents remain in contact with the body. Diarrhea is one way the body has to rid itself of harmful intestinal invaders; it's not a good idea to interfere with this natural response. For that reason, antidiarrheal preparations should be used cautiously, and only after consultation with a physician.

Would an enema help to clear toxins from the intestines?

Using an enema when you have diarrhea is not recommended. The last thing you want to do is create more bowel movements, or softer, more watery stool. In any case, the water of an enema reaches only the last few inches of the large intestine and could not cleanse the rest of the intestinal tract.

If I'm on medication, should I continue it if I'm vomiting or have diarrhea?

This is an important question to which there is no all-purpose answer. It depends on the medication: Only your doctor is in a position to decide whether a particular medication should be discontinued. For example, if you are taking insulin or other medication to control diabetes, it's extremely important that your doctor monitor your blood sugar level to avoid serious complications from developing.

Do all these points apply to all kinds of mild infectious gastroenteritis?

Yes. Gastroenteritis, regardless of the cause, is generally treated the same way: bed rest and fluid and sugar replacement. The infection is usually self-limiting, and the body's natural immune system will take care of the problem by itself.

What are the symptoms of a more serious problem?

Any blood, mucus, or pus in your stool may be a sign of a more severe infection and should be brought to your doctor's attention.

If your eyes feel dry, if your mouth is parched, or if you are not urinating as often as you usually do, you may be dehydrated, and should consult your doctor. Headaches, dizziness, or fainting may also occur with dehydration, so you should be on the alert for these symptoms as well.

When you vomit, check to see what kind of material you bring up. If there's blood in it, or particles that look like coffee grounds, report it to your physician. These signs may indicate bleeding from the upper intestine.

Keep track of your temperature. With mild infectious gastroenteritis, your temperature should not go much above 101°F; a higher temperature may be a sign of a more serious infection and calls for a doctor's examination.

There are many serious diseases and problems that begin in much the same way as mild infectious gastroenteritis. If the symptoms do not go away completely within a day or two, you should see your doctor.

What causes more serious types of gastroenteritis?
Whether an infectious gastroenteritis is mild or severe, it is usually caused by one of three types of organisms: virus, bacterium, or protozoan.

How do organisms enter the intestine?
There is only one way to contract an intestinal infection—by ingesting food or water contaminated with the offending organism. You can't breathe it in like a cold virus, and you can't catch it from being near someone who has it. There must be direct oral contact. Poor hygiene is the major reason these infections spread. For example, if a carrier of an organism doesn't wash his or her hands properly after defecating, material from the contaminated stool may transmit the organism, via food or liquid, to another person. Even if a person has recovered from a bout of infectious gastroenteritis, large amounts of the organism may be passed in the stool. This is referred to as the carrier state.

Infectious gastroenteritis may also be sexually transmitted. This is especially common among homosexuals.

Are there any other sources of an intestinal infection?
It has long been known that dogs may carry a bacterium called *Campylobacter* in their stools. Recently, *Campylobacter* has been identified as a cause of a severe form of infectious gastroenteritis in humans that is often accompanied by a high fever and bloody diarrhea. Outbreaks of the infection have been seen not only in people who have brought young puppies into

their homes, but also in people who had no direct contact with dogs.

Although *Campylobacter* enteritis is usually self-limiting, it sometimes becomes a more serious infection. If high fever, pain, or bloody diarrhea is present, specific treatment with erythromycin or other antibiotics may be required, so a physician should be consulted immediately.

What is food poisoning?

Food poisoning is caused by eating contaminated food. The most common kind of food poisoning is caused by the staphylococcus bacterium. It comes most often from a wound or skin infection of someone who handles food; it thrives in cream sauces, luncheon meats, custards, mayonnaise, and salads made with mayonnaise, especially when these are not kept properly chilled. The dangers are greatest during warm weather. Once the staphylococcus enters these foods, it multiplies rapidly and produces the toxins that are the actual illness-producing agents.

The symptoms of food poisoning don't take long to appear. Two to four hours after eating the contaminated food, nausea, vomiting, diarrhea, and cramps occur. Sometimes, low-grade fever and headaches are also present. No matter how bad the symptoms are, they usually begin to subside within six to twelve hours. Once they are gone, recovery is complete. There is no other bacterial infection that strikes so suddenly or disappears so rapidly.

Salmonella enteritis (or salmonellosis), another type of gastroenteritis, is caused by *Salmonella. Salmonella,* like the staphylococcus, is found in contaminated food, but it may also be present in the stool of people with no symptoms; these people are called carriers. Initially, the symptoms of *Salmonella* poisoning are similar to those of staphyloccocal poisoning, except they don't appear as quickly. Usually, it takes six to forty-eight hours for the first signs of discomfort to occur, and symptoms can last three days or longer.

Though the symptoms seem similar, *Salmonella* poisoning is potentially more serious than staphylococcal poisoning. *Salmonella* may enter the bloodstream and form abscesses in other parts of the body. Fortunately, this is a rare occurrence. Once you have a *Salmonella* infection, you may remain a carrier with the *Salmonella* present in the stool for months after the symptoms disappear.

Another type of bacterial gastroenteritis, shigellosis, is caused by *Shigella*. That pathogen can be ingested from contaminated water as well as food, or can be transmitted by an infected person. The symptoms, which strike abruptly, include fever, vomiting, nausea, cramps, and massive diarrhea (sometimes as frequent as twenty times a day). In healthy adults, the infection usually runs its course. But if the body cannot combat the infection and the disease continues unchecked, *Shigella* infection leads to bacillary dysentery; the lining of the intestine becomes ulcerated, resulting in watery stools filled with pus, mucus, and sometimes, blood.

How is it established whether a person is a carrier?
The carrier state is established by means of a stool analysis and culture that determine whether there are any live organisms present. It is especially important for people who handle food to have this test.

What is traveler's diarrhea?
This is a type of gastroenteritis, or intestinal infection, that is usually self-limited, lasting three to five days. However, for a traveler this may account for an entire vacation, and can be a distressing problem. About 40 percent of people who travel to foreign countries develop a form of watery diarrhea that is referred to by physicians as traveler's diarrhea. Other common names used to describe this entity include "Turista" and "Montezuma's Revenge." It can strike either during a trip or, if exposure occurs on the last day or two of vacation, when the traveler returns home.

Are there other symptoms associated with traveler's diarrhea?

Yes. The diarrhea is often accompanied by cramps, fever, bloody stools, and, sometimes, vomiting.

What causes traveler's diarrhea?

The most frequent cause of traveler's diarrhea is a bacterium called *Escherichia coli* , or *E coli* for short. Unlike the harmless forms of *E coli* you may recall learning about in biology class, traveler's diarrhea is associated with a more toxic strain called enteropathogenic *E coli.* This strain produces a toxin believed to be the cause of the symptoms attributed to traveler's diarrhea.

Are there other causes of traveler's diarrhea?

Any type of gastroenteritis can occur in the traveler, including viral, bacterial (*Salmonella* and *Shigella*, for example), and protozoal (ameba and *Giardia*, for example). However, because enteropathogenic strains of *E coli* are by far the most common cause of intestinal infection in travelers, the term traveler's diarrhea is used to refer to this type of gastroenteritis when the symptoms occur in a traveler.

What can I do to avoid traveler's diarrhea?

Simple precautionary measures are called for, such as never eating raw meats or raw seafood, avoiding dairy products, and avoiding raw fruits and vegetables (or at least peeling them if you absolutely cannot resist a tempting tropical treat). As for drinks, try to stick with beer, wine, hot coffee, hot tea, and carbonated beverages without ice. Whatever you eat or drink, *don't* buy it from a street vendor.

Water is a major source of contamination. If you do drink water, make sure it is bottled water or has been treated with chlorine or iodine. If there is no bottled water in your hotel room, brush your teeth and rinse your mouth with water from the hot water faucet. This is much safer.

The more careful you are, the less likely you are to develop an infection. However, there are no guarantees. Tourists have traveled through foreign countries subsisting on only bottled water and canned food, and have still developed diarrhea, while others have thrown caution to the wind and have not had one sick day.

Should I avoid visiting certain countries?
It would be a shame to allow a fear of traveler's diarrhea to rob you of all the rich experiences that traveling has to offer. I would, however, advise caution, particularly if you have any other health problems. There are certain areas that are considered high risk. Developing countries in Latin America, Africa, the Middle East, and China are the least safe. Traveling in the United States, Canada, Northern Europe, Australia, New Zealand, and many of the Caribbean Islands is the safest.

What can I take to prevent traveler's diarrhea?
Many patients request a prescription for such antidiarrheal medications as Lomotil® (diphenoxylate) or Imodium® (loperamide) to take with them on a trip in hopes of preventing an infection. Neither of these drugs actually helps in prevention, but they can relieve symptoms.

Such medication should be used only in limited quantities because antidiarrheals limit the ability of the bowel to expel the infectious organisms present in the stool. There is a trade-off involved: on the one hand, these medicines enable you to enjoy your vacation with less of the distressing symptoms of crampy pain and diarrhea; on the other hand, they prolong the overall duration of the illness by retaining the infected stool within the bowel.

As for prevention, a medication that may actually keep traveler's diarrhea from developing is Pepto-Bismol®. However, it must be taken, two ounces four times a day, from the day of departure to the day of return. This means either carrying a substantial amount of Pepto-Bismol® with you when you travel, or buying it in the towns or cities you will be visiting. Most travelers find this too inconvenient and choose to take their chances with "Montezuma's Revenge."

Are antibiotics of any value in preventing traveler's diarrhea?

Tetracyclines such as doxycycline (Vibramycin®) can reduce the severity of traveler's diarrhea if they are taken daily. However, photosensitivity may occur with these drugs. (Photosensitivity refers to a diffuse, irritating red rash of the skin on all areas of the body exposed to sunlight.) This can be especially severe in vacationers spending time sunbathing or at the beach. General complications of antibiotics include diarrhea from the drug itself and vaginal yeast infections. Perhaps a safer antibiotic found to be effective in preventing traveler's diarrhea is trimethoprim-sulfamethoxazole (Bactrim™ and Septra®). This medication must be avoided by people allergic to sulfa drugs.

Is it safe to buy medication in a foreign country?

No, not unless you see a doctor in that country. Many countries do not have the stringent drug laws of the United States, and dangerous medications are sold over the counter. Enterovioform is an example of an over-the-counter medication you may see while traveling outside the United States. This drug can cause serious neurologic side effects and should be avoided.

If I get diarrhea, why don't the natives?

When you live in one place for any length of time, your intestinal tract adapts to the native bacterial flora surrounding you. For example, a New Yorker may develop diarrhea when he or she visits Mexico, while a Mexican may develop diarrhea during a visit to New York. With prolonged exposure, the human intestinal tract is able to adapt to most environmental bacteria.

Are some people more likely than others to develop traveler's diarrhea?
It seems to occur more frequently in younger people, perhaps because they are more adventurous in their eating habits and more likely to sample native cuisines.

What should I do if I get traveler's diarrhea?
Traveler's diarrhea, as with any mild gastroenteritis, is self-limiting. However, the dehydration accompanying the diarrhea can create problems of its own if it's not handled properly. Follow the rules for preventing dehydration (see p. 132), try to stay out of the sun while the diarrhea continues, and use antidiarrheal medication sparingly, if at all.

All symptoms of intestinal infection should be watched carefully. If the infection is caused not by an *E coli* bacterium but by a protozoan, and if there is blood or mucus in the stool, the situation is more serious. If you are traveling and develop the symptoms of intestinal infection, stay alert to any complications. If the symptoms are unusually severe or last more than a few days, get in touch with your doctor at home or with a reliable doctor in the country in which you're staying.

What are protozoal infections?
The two most common protozoal infections are giardiasis and amebiasis, both of which can be contracted on a vacation or at home.

What is amebiasis?
Amebiasis is one of the more serious intestinal infections. It is transmitted by contact with contaminated food or water or during sexual intimacy; it is prevalent where human fecal waste is disposed of improperly or where sewage disposal is inadequate. The cause of amebiasis is a protozoan (one-celled animal organism) called *Entamoeba histolytica.* The disease is difficult to control because the infected person does not necessarily have any obvious symptoms. Carriers can be identi-

fied only by a stool analysis for the presence of amebic cysts. Recently, amebiasis has become more common in the United States, especially in metropolitan areas.

Because amebiasis can resemble many other intestinal disorders, diagnosis may take a long time. Ingested as a cyst, the amebas may live in the intestines quietly for weeks or even months, dividing and multiplying rapidly but causing no symptoms, or just vague ones. There may be mild diarrhea, low-grade fever, fatigue, and occasional abdominal pains. Gradually, as the organisms become entrenched within the intestinal walls, the wall begins to ulcerate, causing more severe symptoms. Pus, mucus, and blood may appear in the stool.

The amebas are usually discovered in a stool analysis. Once detected, vigorous medical treatment must begin immediately to prevent them from spreading to other parts of the body, such as the liver. Repeated stool analyses should be performed to make sure all the organisms have been eradicated by the treatment.

How is amebiasis treated?

Amebiasis may be treated with various oral medications available by prescription. The most effective are metronidazole (Flagyl®) and iodoquinol (Yodoxin®). Treatment is usually curative after ten days to two weeks, assuming there has not been reinfection during the treatment period.

What is giardiasis?

Giardiasis is caused by *Giardia lamblia*, another protozoan (one-celled animal organism) like the ameba. The symptoms of giardiasis, which include diarrhea and gaseous distention with foul-smelling stool, are usually not as severe as those of amebiasis. But because the primary site of infection is in the small intestine, it is harder to diagnose. Only half the time will the *Giardia* cysts appear on analysis of a stool specimen.

If your doctor suspects you might have a *Giardia* infection, he or she may confirm the diagnosis by passing a small plastic tube down into the small intestine.

By direct aspiration of the juices from the small intestine through this tube, the microscopic presence of the protozoan may be identified.

How is giardiasis treated?

Metronidazole (Flagyl®) and quinacrine (Atabrine®) are both effective oral treatments for giardiasis.

Are there other considerations?

The symptoms of an intestinal infection are often similar to those of other gastrointestinal problems. Diarrhea, in particular, may be a symptom of many disorders. If you have diarrhea persisting beyond a few days, it's wise to discuss it with your doctor.

16

LIVER DISEASES

Lying in the right upper portion of your abdomen, protected by your rib cage, is the largest and one of the most important organs in your body: the liver. Its functions are so far-reaching that it plays a role in the well-being of almost every system of the body. The immune, endocrine, circulatory, and digestive systems are all dependent upon the liver. As long as the liver remains healthy, all is well; but once this intricate chemical factory begins to break down, the body is thrown into chaos. Liver disease can cause jaundice (yellowing of the skin and whites of the eyes), depress appetite, and cause fatigue and weakness. If the liver is severely damaged, there is the constant risk of infection, abnormalities of the blood-clotting system, and even death.

Many things can interfere with the proper functioning of the liver, including alcohol, medications, environmental toxins, and various bacteria and viruses. One of the most common liver disorders is a viral disease called hepatitis type A. Many adults have been exposed to hepatitis type A as children or adolescents, often without even knowing it, and have developed immunity by adulthood. Though viral hepatitis can be a mild disease, it can also be serious enough to be life-threatening. Knowing what causes it and how to prevent it can sometimes mean the difference between health and disease.

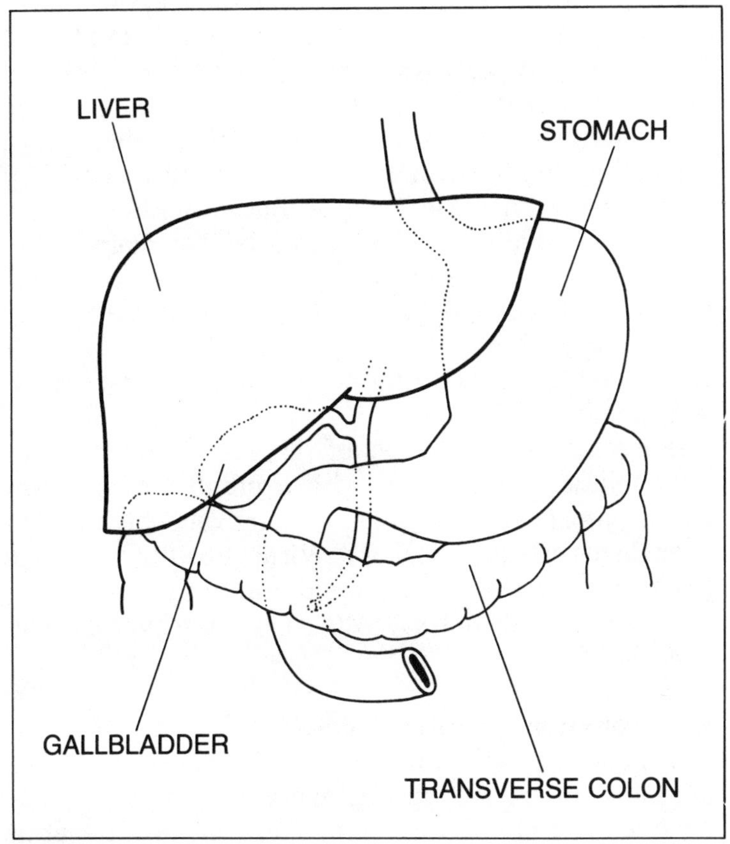

The liver.

What is the liver's function?
The liver is the largest organ in the body, accounting for approximately one-fifth of your total body weight. Unlike all the other digestive organs discussed here, the liver also plays an important role in many other bodily functions.

As a digestive organ, it manufactures the bile necessary for the breakdown and absorption of dietary fats and fat-soluble vitamins A, D, E, and K. It also metabolizes cholesterol and stores glycogen, which is an energy source that fuels the muscles.

The liver, which is normally soft and smooth, is an important weapon against the invasion of the body by bacteria and toxins. It metabolizes or detoxifies alcohol, drugs, and any poisons entering your body. It also filters out and destroys bacteria carried to it by the bloodstream from the intestinal tract.

In addition, the liver acts as a chemical factory, regulating bodily hormones and seeing to it that they do their appointed jobs. It produces blood proteins, such as albumin, that are essential for maintaining plasma volume and transporting enzymes and hormones.

What are the most common diseases affecting the liver's functioning?

The most common disease of the liver is hepatitis, an inflammation of the liver with a number of different causes. Most often, hepatitis results from the invasion of the liver by specific viruses. Viral hepatitis can range from a simple, self-limiting condition to cirrhosis, a disease that actually destroys the cells of the liver and can shorten life.

How common is viral hepatitis?

The incidence of viral hepatitis is increasing at an alarming rate. One out of every two hundred fifty Americans will contract some form of viral hepatitis this year, which means that over 1 million new cases will occur. All indications point to this number increasing over the next five years.

Is there more than one kind of viral hepatitis?

Yes. There are actually many types of viral hepatitis, although only these four have been identified as of this printing: hepatitis A, hepatitis B, hepatitis non-A, non-B, and hepatitis D (Delta). Hepatitis non-A, non-B is really a catchall term for all the forms of hepatitis yet to be identified.

Do all of these viruses produce the same illness?
No. Each type of virus produces a different problem, with different modes of transmission and the possibility of different complications. However, many people think of them as the same because their initial symptoms are usually similar, except in their severity.

What are the symptoms of viral hepatitis?
The severity of symptoms varies with the virus and with the individual person. Usually one or more of the following symptoms occurs: nausea, depressed appetite, pale, loose stools, weakness, fatigue, moderate fever, abdominal pain, joint pain, dark urine, and jaundice. If you have any of these symptoms, get in touch with your doctor immediately. He or she will perform a thorough examination and take blood tests to determine if you do have hepatitis and if serious liver damage has resulted.

Does everyone with hepatitis turn yellow?
No, this is a popular misconception. Only a little more than 10 percent of people with viral hepatitis turn yellow.

What should I do if I have come into contact with a person who has hepatitis?
If you have been in close personal contact with someone who has hepatitis, your doctor may suggest you receive an injection against the virus. "Close personal contact" excludes everyone but spouses, sexual partners, and family members. A casual contact does not necessitate further measures.

If your doctor thinks you may have been exposed to hepatitis A, he or she may recommend a gamma globulin injection to help prevent the symptoms of this form of hepatitis from developing. If the possible exposure was to hepatitis B virus, your physician may give you an immediate injection of hepatitis B immunoglobulin to stave off the disease or to at least lessen its severity.

If your doctor confirms that you have hepatitis, it is

important to tell anyone with whom you have been in close contact (such as spouse, household members, sexual partners) so they can be immunized.

What is hepatitis A?

Hepatitis A, often referred to as infectious hepatitis, is the mildest of all viral hepatitis infections. Often, the symptoms are so mild people aren't even aware they have hepatitis, and just assume they're suffering from an ordinary viral infection. More than 75 percent of people over fifty years old have had hepatitis A, although many don't know it. Once you've had the infection, you acquire a lifelong immunity against this particular strain of virus.

How long is the incubation period for hepatitis A?

Hepatitis A has the shortest incubation period of all the hepatitis viruses. It usually takes from four to six weeks for the symptoms to develop after the initial contact.

How is hepatitis A spread?

Hepatitis A is an oral-fecal disease, which means that its mode of transmission is through contact with fecal material contaminated with the hepatitis A virus. This may happen directly, through sexual contact or, more commonly, indirectly, by infected individuals not washing their hands properly after a bowel movement. Even a minute amount of fecal material on the hands can be carried into food or onto eating utensils.

What is hepatitis B?

Hepatitis B is a more serious form of viral hepatitis.

How is it spread?

Until recently, hepatitis B was called serum hepatitis because its most common form of transmission was through contaminated blood transfusions. Now there's a way to screen blood for the presence of the virus, so it is no longer transmitted that way. However, it can still be passed on through dental and medical procedures if the instruments are not sterilized or disposed of properly. Tattooing, acupuncture, ear-piercing, intravenous drug abuse, and accidental needle pricks in hospitals and on kidney dialysis units are examples of direct routes of transmitting hepatitis B.

But the most common form of transmission is through sexual contact, for the hepatitis B virus lives in the body fluids of an infected individual and can be passed on through saliva and semen. It is much less likely to be transmitted through casual contact, and is not thought to be present in feces.

Are the symptoms the same as with hepatitis A?
Yes but they are usually more severe and last for a longer period of time. Up to 20 percent of people with hepatitis B still report symptoms and have positive blood tests six months after the onset of symptoms.

Even after the symptoms have subsided, the person can still be a carrier of the virus. The virus will stay alive in the body until the antibodies the body produces fight it off. This can take years. To date, there is no cure for the hepatitis B carrier state, and individuals become a reservoir for its transmission. About one in every five hundred people in the United States carries this virus.

How can a carrier of the hepatitis B virus avoid spreading the disease?
The only thing a carrier can do is exercise caution in his or her contacts with other people and, of course, never donate blood.

Can I develop immunity against the hepatitis B virus once I've had hepatitis B?
In order to develop an immunity against the hepatitis B virus, your body must first build up the appropriate antibodies against it. Sometimes this can take months or even years.

How do I know when I have developed the antibodies against the disease?
Your doctor can determine this with a blood test.

What is the incubation period for hepatitis B?
Hepatitis B has a long incubation period. After exposure to the virus, it can take anywhere from forty days to six months for symptoms to develop, with the average being about two months. This extended incubation period often makes it difficult to pinpoint the original source of infection. Hepatitis B is most contagious before the development of symptoms.

Is there any way to avoid hepatitis B?
Yes. A vaccine is now available that offers lifelong immunity against the hepatitis B virus. Anyone who has come into contact with an infected person, as well as those who are at special risk of developing the disease, should be immunized. The vaccine will be especially valuable to surgeons, dentists, people working in dialysis units, people on dialysis, and homosexual men; these groups seem to have the highest incidence of the disease.

What is Delta hepatitis?
Delta hepatitis, or hepatitis D, was first recognized in 1985, so researchers are still studying this strain. So far, it seems that the Delta virus is only a problem to people who already have hepatitis B, since it coexists with that virus but cannot live on its own.

How is Delta hepatitis spread?
Much like the hepatitis B virus, it is spread by blood transfusions, unsterile needles, and sexual contact. It has been isolated predominantly in homosexual men.

What is non-A, non-B hepatitis?
It is the collective name for all the viruses that may cause hepatitis but are neither the A, B, nor Delta strains.

How is non-A, non-B hepatitis spread?
It is spread almost exclusively as a result of blood transfusions. There is as yet no method to screen donor blood for non-A, non-B hepatitis.

What are the symptoms of non-A, non-B hepatitis?
The symptoms may be any or all of the typical hepatitis symptoms but, in general, they are mild.

Is non-A, non-B hepatitis a serious form of the virus?
Yes, it can be. Even though the symptoms are mild, these viruses have the the potential for causing more serious liver complications, including chronic hepatitis and cirrhosis.

What is the incubation period for non-A, non-B hepatitis?
The incubation period can be anywhere from two to ten weeks.

How is hepatitis treated?
All types of hepatitis are generally treated in the same way. Since there is no specific medication for the treatment of hepatitis, the person is advised to take it easy and rest until the symptoms have abated. Eating a well-balanced, nutritious diet during this period is particularly important, as the body requires extra energy and nutrients for defense against the infection. Your doctor will continue to take periodic blood samples to check on the course of the disease and to see that no further complications set in.

What complications could occur?
Occasionally, viral hepatitis interferes with normal blood-clotting mechanisms; if this happens, treatment with vitamin K becomes necessary. With hepatitis B and hepatitis non-A, non-B, there is a chance of developing chronic hepatitis, which may lead to cirrhosis of the liver. About 10 percent of people with these viral

infections develop cirrhosis of the liver. This never occurs with hepatitis A.

What is cirrhosis?

In cirrhosis, normal liver cells are damaged and replaced by scar tissue. The scarring that results interferes with the normal functioning of the liver and may cause serious problems. Cirrhosis can be caused not only by hepatitis, but also by excessive use of alcohol or drugs, prolonged exposure to environmental toxins, and rare, inherited forms of liver disease. Once cirrhosis occurs, it is irreversible.

How serious is cirrhosis?

Cirrhosis causes thirty thousand deaths in this country each year, making it one of the four leading causes of mortality. Anything that compromises the flow of blood to and from the liver, such as scarring, is a serious problem. For example, a well-functioning liver is necessary to detoxify any drugs or alcohol ingested; when the liver is not working up to par, these ordinary chemicals may become toxic to the system.

Ammonia, a by-product of protein digestion, may build up in the blood when the liver is no longer able to detoxify it properly. This ammonia then enters the nervous system and can contribute to severe mental changes; these, in turn, can eventually lead to hepatic coma and death.

The liver is a giant lymphoid organ (part of the immune system) with highly specialized cells that destroy invading bacteria. If its immune abilities are compromised, almost any infection can become life-threatening. The liver produces albumin, an essential protein that helps maintain the blood plasma level within the blood vessels. If the albumin level falls, plasma fluids may leak out of the blood vessels and through the surface of the liver into the abdominal cavity. This leakage produces abdominal swelling, called ascites, and may also cause swelling in other tissues.

A large amount of scarring increases the resistance

to the flow of blood into the liver. This eventually results in congestion and engorgement of the veins that drain into the liver, including those that line the inside of the esophagus. This condition is called esophageal varices (varix means an abnormally distended vein). When such varices develop, they can tear, causing life-threatening hemorrhage.

Are there further considerations?

Yes. During the period of active hepatitis infection, it is important that the liver not be abused or overworked. Alcohol and certain drugs place an added strain on the liver, so all alcohol should be avoided during this period, and you should check with your doctor about taking any medications. Certain drugs, including tranquilizers, some tetracyclines and other antibiotics, antidepressants, and excessive acetaminophen (Tylenol®, Datril®) are best avoided while the liver is inflamed.

17
CANCER

Almost everyone has been touched by cancer in some way. It strikes one out of four people, and two out of three families. This year alone, it is estimated that 800,000 new cases will be diagnosed, and about 200,000 of them will involve the digestive tract. Except for lung cancer, bowel cancer is the most common type of cancer in this country.

The statistics are staggering and, until recently, the word cancer meant only death. But now, with improved diagnostic tools allowing for early detection of many malignancies and promising new developments in medical and surgical treatment, the outlook for cancer patients is far more optimistic. Life expectancy of cancer victims has been extended, and the word "cure" is now a reality in some cases.

Cancer has become a household word, yet few people really know what it means. Actually, cancer is not one disease, but a general term covering more than one hundred separate diseases, all of which have one common denominator. In each, there is an abnormal and uncontrolled growth of cells resulting in the development of a mass called a tumor. Cells from these tumors can invade and destroy surrounding tissue and spread through the bloodstream or the lymph vessels to produce new tumors, called metastases, in other areas of the body.

The study of cancer has its own terminology; in order to understand the different forms of the disease, as well as the diagnostic tools and modes of treatment, one must become familiar with it. But even more important is knowing what the early symptoms of these diseases are and what symptoms you should bring to your doctor's attention. This chapter will describe the types of cancer that can attack each part of the digestive tract, from the esophagus to the rectum. In cancer, early detection can often make the difference between life and death.

What are some of the terms associated with cancer?
The following list explains some of the common terms.

- **Benign tumor:** Any growth that is not cancerous and does not spread to other parts of the body.
- **Biopsy:** The removal and microscopic examination of a tissue sample, either benign or malignant.
- **Chemotherapy:** Treatment with anticancer chemical agents that are either injected into the bloodstream or taken orally. These chemicals are intended to destroy cancer cells with as little damage as possible to normal cells.
- **Malignant tumor:** Cancerous tumor.
- **Metastasis:** A secondary cancer growth resulting from the transfer of cells from a primary malignant tumor to another part of the body.
- **Oncologist:** A doctor who is specially trained in the treatment of cancer and who administers anticancer therapy.
- **Radiation therapy:** Treatment of cancer using x-rays or radioactive material.
- **Tumor:** Any abnormal growth of cells, benign or malignant.

What are the symptoms of cancer of the esophagus?
The most common symptom of cancer of the esophagus is dysphagia, or difficulty in swallowing. The sensation is of food getting stuck along the way as you swallow it. Solid foods, such as bread and meat, are usually the first foods to cause this sticking sensation. There may also be a burning pain, as food is swallowed, that seems to arise from behind the breastbone in the center of the chest.

Who is most likely to develop cancer of the esophagus?
Studies have been done to determine whether certain characteristics can be linked with esophageal cancer. The disease is more prevalent in men than in women and in blacks than in whites. It is more common in Japan than in the United States; researchers are studying the Japanese diet and life-style to see if they can find the reason for this.

Some researchers think esophageal cancer is linked to heavy consumption of alcohol and cigarettes, but not all people who have it drink or smoke.

Studies seem to indicate that, unlike some other types of cancer, it does not run in families.

How is esophageal cancer diagnosed?
If you have any symptoms I have mentioned, your doctor may recommend tests. The initial one may be an upper gastrointestinal x-ray film series, or UGI. You will be asked to swallow barium, a radiopaque dye, to help make your esophagus visible on an x-ray examination. In addition to taking regular x-rays films, the radiologist will view you on a fluoroscope screen. If any abnormalities are seen on this x-ray examination, your physician may recommend you have flexible esophagoscopy. This is a simple procedure that can be done in the doctor's office or in the outpatient department of a hospital. After giving you a local anesthetic, your doctor (usually a gastroenterologist) will pass a narrow,

flexible fiberoptic tube into your mouth and down your esophagus. Through the tube, he or she can take a tissue sample for a biopsy and photograph certain areas to be examined later on.

How is esophageal cancer treated?

If you are found to have esophageal cancer, your doctor will take many factors into consideration in deciding the best course of treatment for you, including your general health, the type and extent of the tumor, and the part of the esophagus in which it is located.

Generally, most tumors originating in the esophagus are squamous-cell carcinomas. (Squamous refers to the cells that line the inside surface of the esophagus.) This type of tumor is sensitive to radiation. For this reason, radiotherapy is used in conjunction with surgery in an attempt to cure primary cancer of the esophagus.

Many tumors of the esophagus actually originate in the stomach. They are called adenocarcinomas and are not as sensitive to radiation as squamous-cell carcinomas. Therefore, early surgery is relied upon for definitive (or curative) treatment.

Does surgery for cancer of the esophagus require removal of part of the esophagus?

Yes. Removing a malignant tumor of the esophagus generally involves removing that segment of the esophagus that is involved with the cancer. However, the remaining esophagus is usually long enough for the surgeon to reestablish continuity and maintain a conduit for the normal passage of food. Unfortunately, however, this is not always the case. When the tumor is advanced and widespread the surgeon must remove most of the esophagus. In this situation, a short segment of the patient's colon is removed and interposed to replace the resected esophagus. As an alternative, when surgical cure is not possible, the surgeon may implant a prosthetic tube directly into the esophagus to maintain patency and allow for normal swallowing.

Are there alternatives to esophageal surgery?
Yes. An alternative to surgery is laser therapy used in conjunction with a gastroscope. A gastroscope is a soft, flexible tube with a narrow diameter. It is filled with fiberoptic bundles that project light. It is passed through the mouth into the esophagus and can be advanced into the stomach and duodenum, illuminating the interior of these organs. The laser is a form of amplified light that may be directed through a gastroscope and applied directly to the surface of the tumor to reduce tumor mass. However, lasers are palliative, not curative. They are used only for recurrent tumors or in the presence of widespread disease that is too far advanced to establish a primary cure with surgery. The risks of laser treatment, which include perforation and bleeding, may be life-threatening in patients with advanced cancer.

How prevalent is stomach cancer?
It is estimated that approximately 24,000 new cases of stomach cancer will be diagnosed each year.

Are some people predisposed to developing stomach cancer?
Yes. Research has demonstrated that certain characteristics are often associated with stomach cancer. Males with stomach cancer outnumber females three to one. Stomach cancer is more likely to occur in people over the age of fifty and in people with blood type A. It is also more common among the Japanese, Norwegians, and Icelanders.

A condition called atrophic gastritis may indirectly predispose a person to cancer of the stomach. Diabetes mellitus, pernicious anemia, and even the aging process itself all predispose people to atrophic gastritis. This is characterized by the absence of stomach acid, which may in some way be related to the development of stomach cancer.

What are the symptoms of stomach cancer?

The first symptoms of stomach cancer can be much like those of other digestive disorders. There may be persistent indigestion, a bloated or distended feeling, especially after eating, nausea, loss of appetite, heartburn, mild stomach pain, or blood in the stool.

Usually, though, there is a gnawing, persistent pain in the stomach that seems to get worse after eating. This is just the opposite of ulcer pain, which usually abates after meals.

Sometimes there may be no symptoms at all until an obstruction occurs; at this point, the cancer has either grown so large, or is so strategically located, that it obstructs the opening of the stomach into the small intestine.

How is stomach cancer diagnosed?

If your doctor suspects you might have stomach cancer, the first step would be to conduct a complete physical examination with blood tests and stool tests for the presence of blood. Next, your doctor may recommend that you have a UGI x-ray film series. The x-ray examination allows the contour of the stomach to be seen, including any abnormality. If any abnormality is found, your doctor may suggest you have a gastroscopy. This is a simple test that can be done in the physician's office. After administering a sedative and local anesthetic, the doctor introduces a flexible fiberoptic tube through your mouth and into your stomach. This enables the physician to see the inside of the stomach, as well as to take tissue for a biopsy and take photographs for later study.

How is a malignancy in the stomach treated?

Generally, stomach cancer is treated by surgical removal of the malignancy. After surgery, a decision is made as to whether chemotherapy is necessary or advisable. Radiation therapy is not used as often as chemotherapy in these instances because it is less successful against stomach cancer than against other types of cancer, such as lymphoma.

How much of the stomach can safely be removed without serious side effects?

Actually, the stomach is not a vital digestive organ. Its main function in digestion is the storage of a meal; little of the stomach is necessary to perform this function. In fact, you can live without any stomach at all, but meals must be smaller and thus more frequent. In the past fifty years, the rate of stomach cancer has been declining, while other forms of cancer of the digestive tract have been on the rise. Although many theories have been proposed to explain this, no one really knows what is responsible for this downward trend.

Is cancer of the small intestine a common disease?

No. It is actually among the rarest of all digestive cancers. Usually, cancers occurring in the small intestine are secondary metastases from tumors in other parts of the body.

What are the symptoms of cancer of the small intestine?

There may be abdominal pain that cannot be explained, bleeding, diarrhea, or, in rare cases, an obstruction causing acute pain, nausea, and vomiting. A malignancy may grow in such a way that it obstructs one of the openings into or out of the small intestine. If this happens, acute pain may require rapid surgical intervention.

How is malignancy in the small intestine treated?

Again, treatment depends on a person's general health as well as the type and extent of the malignancy. Usually, the only treatment for a primary malignancy of the small intestine is surgery. But you cannot survive without a small intestine, so only portions of it can be removed.

Is liver cancer also rare?

A primary cancer of the liver, like a primary cancer of

the small intestine, is rare. Most cancers growing in the liver are really metastases from cancers in other organs of the body.

Are certain people predisposed to developing primary liver cancer?

Yes. Most primary liver cancers occur where there is preexisting cirrhosis of the liver. Also, there may be an association between liver cancer and the presence of the hepatitis B antigen in the blood; this is especially true in people who have underlying alcoholic liver disease.

Aflatoxin, a fungus that grows on shelled peanuts, may be carcinogenic to the liver. The incidence of primary liver cancer is high in a number of underdeveloped countries where this toxin is prevalent.

What are the symptoms of primary liver cancer?

The symptoms of liver cancer, especially in the early stages, can be quite indefinite. There may be pain in the area of the liver, depressed appetite, weight loss, and fatigue. Later on, there may be dark urine and loose, clay-colored stools.

How is primary liver cancer diagnosed?

If your doctor suspects you might have primary liver cancer, a complete physical examination will be performed, including liver chemistries (blood tests) and an examination of the liver to see if it is enlarged and tender. A special blood test, called the alpha fetoprotein test, may also be done. This test measures a protein in the blood that represents a tumor marker—that is, it may indicate the presence of hepatoma (a liver tumor). This blood test is positive in the majority of people who have hepatoma. However, a negative test does not rule out the presence of hepatoma, and it may be positive in diseases other than liver cancer.

If any abnormalities are indicated by these tests, your physician may want you to have a liver scan, which is a test that shows the outline of the liver and may indicate the presence of a tumor. Your doctor may also perform a liver biopsy. To do this, a needle is inserted through your skin, and liver tissue is removed through the needle. The tissue is then examined under the microscope by a pathologist to see if any malignant cells are present.

How is a malignancy of the liver treated?
As with any malignancy, treatment depends on the age and general health of the person, and the type and extent of the cancer. Usually surgery is performed only if the cancer is limited to a localized area of the liver and there is no underlying cirrhosis.

Most often, chemotherapy is the treatment of choice for liver cancer. An anticancer drug may be injected directly into a vein leading to the liver. During chemotherapy treatments, blood is constantly monitored to check the progress of the disease and to observe any complications that may result from the chemotherapy.

Is cancer of the pancreas also rare?
There seems to be an increase in pancreatic cancer in this country, and it is now considered one of the ten leading causes of death from cancer. This year alone, it is estimated that 25,000 new cases will be diagnosed. This number is particularly significant because cancer of the pancreas is difficult to diagnose and treat.

How is the diagnosis of pancreatic cancer made?
No technique has been developed to completely visualize the inside and the outside of the pancreas. But there are ways to visualize the surrounding tissues and organs, as well as the pancreatic duct. The first test your doctor may suggest is a UGI series. Though the radiologist will not be able to see the pancreas itself, distortion of the small intestine that could be caused by a pancreatic disorder would be visible.

Ultrasound scanning is another device used to help visualize the area. Sound waves are beamed into the area overlaying the pancreas, and the echoes produced are picked up by a recording instrument and fed into a computer that translates the various echoes into images representing intraabdominal organs.

A computerized axial tomogram, or CAT scan, may be done to visualize the area in and around the pancreas. In this case, the computer projects a more detailed image of the abdominal cavity than do conventional x-ray films, and the process takes only a few minutes. Another diagnostic tool, endoscopic retrograde pancreatography, employs a flexible fiberoptic duodenoscope. This instrument is passed through the mouth and enables the injection of dye directly into the main pancreatic duct. By means of this dye, the pancreatic duct can be visualized by x-ray examination.

Though all of these diagnostic tools can be helpful, a conclusive diagnosis can best be made from a tissue biopsy. A thin needle, guided by ultrasound waves or CAT scanning, is passed through the skin into the pancreas. Tissue is removed through the needle and studied under the microscope to see if any malignant cells are present.

What are the symptoms of pancreatic cancer?
Unfortunately, by the time symptoms develop, the cancer is often widespread. The symptoms vary depending on the part or parts of the pancreas involved. There may be acute pain, similar to that caused by pancreatitis, or if the bile duct is involved, there may be symptoms much like those of a gallbladder attack. Because the tip of the pancreas wraps around the bile duct, a malignancy there may cause obstruction, just as a gallstone would. It this happens, there may be jaundice (yellowing of the skin and the whites of the eyes), abdominal pain, swelling of the abdomen, nausea, loss of appetite, fever, and general malaise. However, if the malignancy is in the tail portion of the pancreas (the part that is distant from the bile duct), there may be no symptoms until the cancer is far advanced.

Is there any treatment for pancreatic cancer?
A number of treatments have been tried, but so far none seem to be successful. In rare cases, the pancreas can be removed surgically and diet can be supplemented with digestive enzymes, taken orally with meals, to replace the pancreatic enzymes and insulin injections given to replace the insulin normally secreted by the pancreas. With this surgery, a part of the stomach and the duodenum must also be removed.

Sometimes palliative surgery is implemented to relieve some of the symptoms of jaundice that develop with pancreatic cancer. A bypass around the common bile duct is done so that the bile flows from the liver into the small intestine at a point beyond the area of obstruction.

Is the prognosis for colorectal (bowel) cancer as poor as that for pancreatic cancer?
Colorectal cancer (cancer of the colon and/or rectum) is quite prevalent; approximately 123,000 new cases are being diagnosed each year. But the outlook for those who do develop it is far from grim. New methods of diagnosis and treatment give reason for optimism.

What are the symptoms of colorectal cancer?
The first symptom is usually blood in the stool, though not enough to be noticed on toilet tissue or in the toilet bowl. It is usually discovered when the physician does a routine stool test that can show the presence of occult blood (blood that is invisible to the eye).

Later on, the symptoms become more noticeable. There may be a change in bowel habit, including unexplained constipation or diarrhea, or change in the color, frequency, or consistency of stool. Because of this, any change in bowel habit or any noticeable bleeding from the rectum should be reported immediately to the doctor.

Vitamin C may cause the test for blood in the stool to be falsely negative. This means that if you are taking vitamin C supplements before the stool test, the test may not show the presence of blood in the stool. Consequently, it is a good idea to stop taking vitamin C supplements at least several days before the test is done.

Many foods can cause the test to show a falsely positive result. Meat, especially when undercooked, can make this happen. Prior to taking the test, you should be on a high-fiber diet because fiber tends to thoroughly empty the lower bowel. Therefore, if there is any potential for bleeding from the colon, the blood will appear in the stool sample.

What causes colorectal cancer?

More is known about the causes of colorectal cancer than of malignancies in the other digestive organs. This facilitates early diagnosis and effective treatment. Except in cases of long-standing ulcerative colitis, most colorectal cancers begin within a polyp. Doctors screen for the presence of polyps and remove them as soon as they are discovered.

Nowadays, polyps do not require surgery as they can easily be removed through a colonoscope (see Chapter 13), especially if these polyps are pedunculated (on a stalk). Pedunculated polyps have a tendency to grow in groups. Therefore, once one has been removed, a doctor must be on the alert for the development of new polyps anywhere in the colon. This information has helped not only to prevent colorectal cancer, but also to discover and treat it before it has a chance to metastasize.

A profile has been developed of those groups who are most likely to develop colorectal cancer. People who fit into any of the groups listed here should be especially alert to early symptoms:

- Anyone with a prior history of colorectal cancer or of a polyp in the colon has a greater chance of developing another polyp and malignancy.
- Anyone who has had ulcerative colitis involving

most of the colon for ten years or more is prone to developing cancer in the colon.

- Anyone with hereditary familial polyposis almost always develops colon cancer. Therefore, the colon is surgically removed before any malignancy occurs and all family members and relatives are screened for polyps (see Chapter 13).
- Anyone who has a strong family history of colorectal cancer should have frequent examinations and be particularly observant of potential symptoms.

How is bowel cancer diagnosed?

If you test positive for occult blood in stool and/or have any significant change in your bowel habit, your doctor may recommend that you have further diagnostic tests. The first step is flexible sigmoidscopy. This test employs a short flexible endoscope making it possible to see the lower portions of the colon and rectum. A tissue biopsy can be taken, if necessary.

Your doctor may also request that you have a barium enema. Here, radiopaque dye is introduced into the rectum through an enema tube. This enables any small tumors or other abnormalities of the colon to be visualized by x-ray film.

In addition, your physician may recommend colonoscopy be performed so the entire colon can be visualized. This is a simple procedure that can be done in the office. After giving you a mild sedative, your doctor will introduce a colonoscope (a flexible fiberoptic tube) through your rectum into your colon. Through this tube a biopsy of tissue further along the colon can be taken, as can photographs to be studied later. Also, any polyps detected during the examination can be removed.

Is there a blood test that can determine whether a person has colorectal cancer?

There is a test for the presence of certain types of antigens in the blood known as the carcinoembryonic antigen or CEA for short. CEA has been found in people with colon cancer, but is also present in other diseases, and is associated with cigarette smoking.

The CEA test may serve an important function. When someone is operated on for colorectal cancer, a CEA level is measured both before and after surgery. Because the test results are highest before surgery and lowest after surgery, it provides a clue to the likelihood that another colorectal cancer has developed. In many cases, this test is repeated at yearly intervals after surgery; when used in conjunction with other tests, it can alert the physician to a possible recurrence.

Are all colorectal cancers the same?

No. The most common cancer growing in the colon is called adenocarcinoma. It responds differently to chemotherapy and radiation than the type growing in the lower anorectal area. Anorectal cancers are usually squamous-cell carcinomas and are more sensitive to chemotherapy and radiation.

How are colorectal cancers treated?

Each type is treated differently. The adenocarcinoma is surgically removed whenever possible, leaving the remaining portion of the colon intact. Because adenocarcinoma is insensitive to chemotherapy and radiation, these treatment modalities are often not used.

The squamous-cell carcinomas occur so low in the rectal area that the construction of a permanent colostomy is often necessary when surgery is performed (see Chapter 18). However, these tumors are sensitive to radiation and chemotherapy, both of which are often used in conjunction with surgery.

Can a laser be used to treat colorectal tumors?

Yes. A laser is a form of amplified light now being utilized to treat all types of recurrent colorectal tumors. The laser is projected through the colonoscope and aimed directly at the tumor. This results in destruction of the bulk of the tumor by a process called photocoagulation. Despite its usefulness, laser therapy is not a substitute for primary surgical resection. It is palliative, not curative, and is used only to relieve obstruction or bleeding problems resulting from a recurrence of a colon tumor. It involves such risks as perforation of the wall of the intestine.

18 OSTOMIES

Throughout the years, many medical advances have been made on the battlefield. It is there that doctors, faced with catastrophic injuries, must use their inventiveness to find new ways to save lives.

The colostomy, the forerunner of all ostomies, was first performed by a physician during the American Revolution. In the attempt to save the life of a soldier wounded in the intestine, he removed part of the colon and brought the remaining end of the small intestine outside the body. In this granddaddy of ostomies, the stoma (the opening) was just below the spine. The operation was attempted several times afterwards, but none of the soldiers lived long enough to find out how difficult it would be to care for an ostomy located in the rear portion of the body.

More than a century later, in 1908, Dr. William Ernest Miles used the information obtained from these primitive attempts and successfully performed the first colostomy with the stoma on the outside of the abdominal wall.

The twentieth century has seen many advances in this procedure. Now, people with ostomies live out their normal life span in ease and comfort. They can work, swim, participate in athletics, and continue to enjoy sexual relations.

This chapter is directed to both the person who is facing the procedure and the individual who has recently undergone it. Besides providing the necessary information, it will hopefully put all fears and uncertainties to rest. Having an ostomy is not the end but the beginning—the beginning of a healthier life.

What is an ostomy?
An ostomy is any surgically constructed opening outside the body allowing the passage of wastes.

Why are ostomies performed?
Ostomies are done when the normal intestinal organs of excretion either have to be removed entirely, or temporarily disconnected to allow specific parts of the bowel to heal.

How many kinds of ostomies are there?
There are three types, but because one, the urostomy, is a urological procedure, I will confine this discussion to the colostomy and ileostomy, both of which are gastrointestinal procedures.

The most common ostomy is the colostomy. A major portion of the colon is either permanently removed or temporarily disconnected, leaving a small, remaining segment to be brought through the skin of the abdominal wall.

The ileostomy, a somewhat similar procedure, is performed when the entire colon must be removed or temporarily disconnected. If this is necessary, the end of the small intestine—the ileum—is brought through the abdominal wall.

Another procedure, known as the continent ostomy, is really a variation of the ileostomy. The colon is entirely removed and the ileum is surgically attached to an internal pouch made of intestinal tissue. A tube, made from the same intestinal tissue, extends from the pouch to the outside of the abdominal wall through the skin. Because the pouch is internal, individuals with a continent ileostomy do not have to wear an outside appliance. They control the time and place of evacuation.

Are ostomies common?
Ostomies are far more common than you may realize. As a matter of fact, more ostomies are performed than mastectomies (the surgical removal of the breast). Statistically, more than 100,000 ostomies are done each year—a number that is on the rise as the incidence of colorectal cancer increases.

Why are colostomies so much more common in people over fifty?
This age group experiences the highest incidence of colorectal cancer. When cancer involves the rectum or the lowest portion of the colon near the rectum, the surgeon must create a colostomy in order to remove the entire cancer and maintain a conduit for evacuation. Fortunately, most colon cancers are located well above the rectum, and may be surgically removed without necessitating a colostomy.

If my doctor has recommended that I have an ostomy, how can I be sure I have a competent surgeon?
If you have been advised to have surgery, it is always wise to get a second opinion. You can check the credentials of any physician with your local medical society as well as the hospital that the surgeon is affiliated with. In addition, you can contact one of the local organizations that deal specifically with people who have your disease. (The national organizations listed on p. 177 may be able to provide you with the addresses and telephone numbers of local affiliates.) These organizations are often able to offer helpful advice concerning competent surgeons for your problem.

Is there much pain after the surgery?
Although the ostomy usually heals quickly, the first

few days after the operation can be quite uncomfortable. In most cases, though, there is little actual pain involved, and you can resume your normal activities within about six weeks of the surgery.

What is an enterostomal therapist?
An enterostomal therapist, or ET, for short, is a person specifically trained in ostomy care. Often, an ET will be called in prior to surgery to advise and reassure the person about to undergo the procedure. Postoperatively, they help the person deal both psychologically and physically with any problems that arise. Some ETs have themselves had ostomies, and can offer both their knowledge and practical experience in helping the individual make a quick adjustment.

Most large hospitals have at least one ET on staff. Ask your doctor about it before your surgery. If your hospital does not have an ET directly connected to it, then contact the United Ostomy Association in your area for a referral.

Is it difficult to adjust to an ostomy?
Initially, there may be some problems in adjusting. Common reactions include depression, anger, fear, as well as a change in self-image. Also, there can be a strong sense of the loss of an important part of yourself. Fortunately, none of these negative feelings usually lasts too long. With the help of caring, supportive people, these individuals soon realize that rather than having been robbed of something precious, they have been given something valuable. Along with the ostomy comes a new chance at life. They are not victims; they are survivors.

How much does an ostomy change a person's life?
Everything that happens to us changes us in some way. As we grow older, we are continually faced with new challenges to meet and new obstacles to overcome. It is in coping that we discover strengths and abilities we may not have realized we have. Having an ostomy is

just being challenged in a new way. Life-style changes only as much as you allow it to. In actuality, people with ostomies can swim, fly planes, dance, climb mountains, work, play, and continue enjoying their sexuality to the fullest.

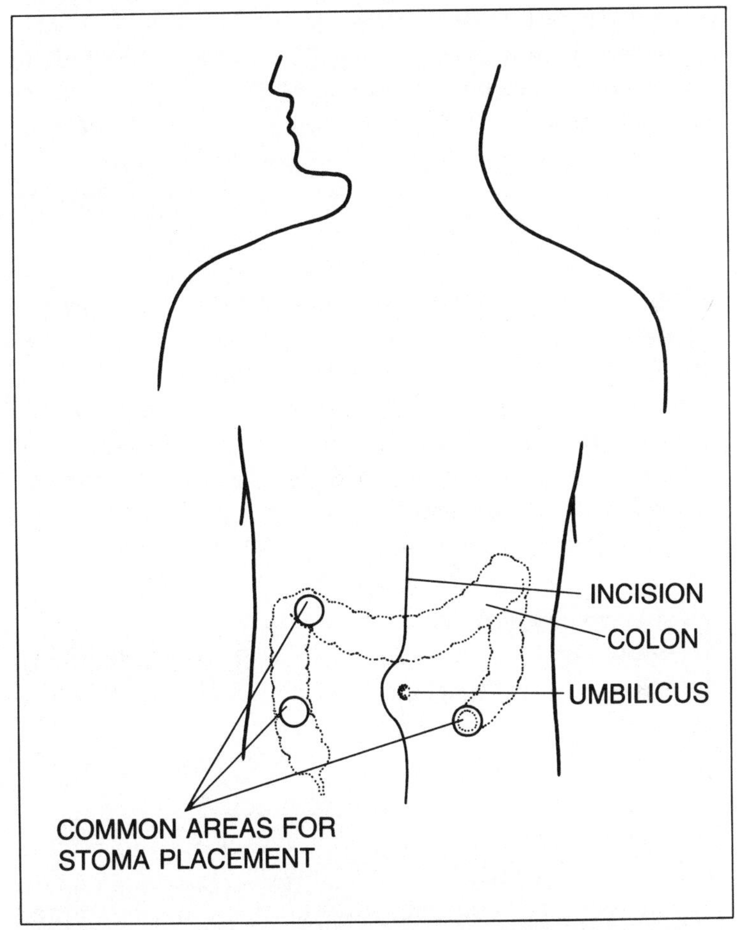

Most common areas for placement of stoma in ostomy procedures.

Could you tell me more about the stoma?

The stoma is the end portion of the colon or small intestine that is brought outside the abdominal wall. No two stomas are alike. They come in all shapes and sizes, ranging from the size of a dime to a span of several inches. The amount of protrusion from the body varies as well. Much depends on the type of ostomy performed, the shape of the intestine, and how much of the intestine the doctor must extend outside the body.

Most people who recently have had the surgery are in awe of their stomas and feel they have to be carefully protected. Of course, a certain amount of caution is always advised, but stomas are resilient. I have never heard of one harmed by a shower, a bump, or sexual activity. On occasion, there might be some blood oozing out of a stoma, but this is usually because intestinal tissue has an extremely rich blood supply, not because the stoma has been injured.

Does a normal bowel movement pass through the stoma?

One of the main functions of the ileum and colon is to remove excess water from the stool. When these organs are no longer functioning, the stool is watery. A person with an ostomy can expect to have loose and watery stools, and should not consider this to be a sign of a problem.

What is an appliance?

An appliance is a thin, flexible pouch that adheres to the skin around the stoma and acts as a waste collector.

Are there different kinds of appliances?

Yes. There are basically two kinds: disposable ones, which are discarded after a few days; and reusable ones, which are cleansed thoroughly and used over and over again. The disposables are much easier to handle but are more expensive. Both types can be used either alone or with a belt that adds extra support during athletics or sexual activity.

Is odor a problem?

It has been in the past. But now, appliances are fitted more securely, and deodorants are available that can

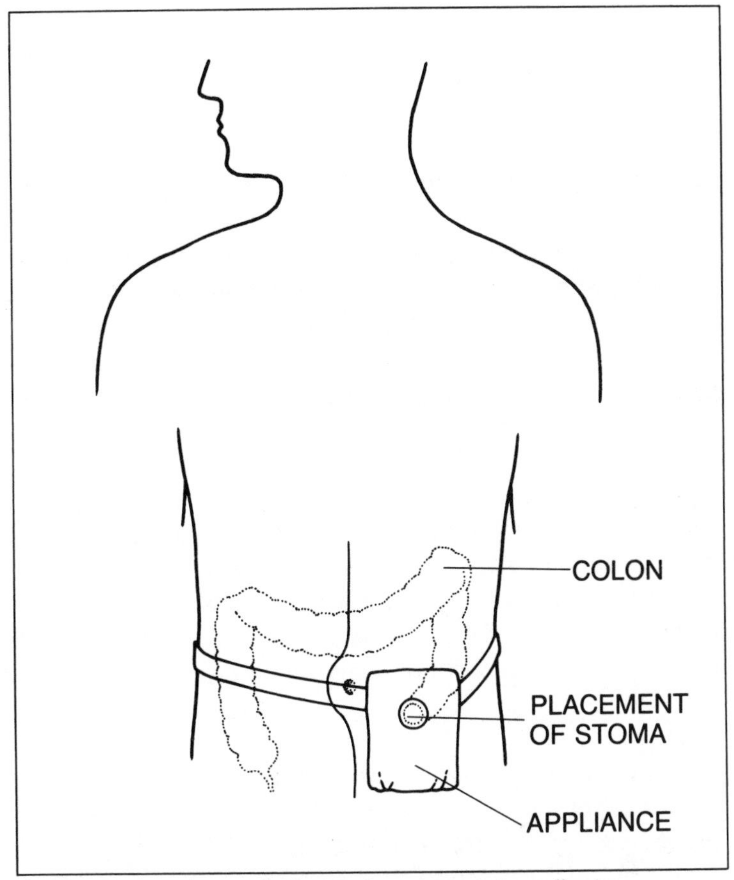

An appliance, a pouch that acts as a waste collector.

be added to the appliances to eliminate most of the odor.

Do people with ostomies have to be on special medications?

As a general rule, no. But there are times when medication is needed to help control diarrhea, particularly if a viral illness develops or if the individual is taking antibiotics. Some people who have had this surgery find they have an occasional problem with constipation. This is usually due to a faulty diet, lack of exercise, or inadequate fluid intake. Whether the problem is diarrhea or constipation, the physician should be notified. A person with an ostomy should never self-medicate or use any over-the-counter preparation without first speaking to a doctor. The wrong medication can cause serious problems.

Do people with ostomies have to stay on special diets?
Some individuals can eat anything they want without experiencing irritation or a change in bowel habits. Others find certain foods increase the amount of gas or change the consistency of their stools. If there is a problem, my advice is to go on a bland diet and add one new food a day to determine which is causing the difficulty.

Are enemas appropriate for people with ostomies?
An ordinary enema, that is, one placed in the anus, is never used. However, there are times when the stoma has to be irrigated to facilitate the passage of stool. Of course, this should only be done on the advice of a physician.

How can I obtain more information on ostomies?
First address any questions you might have to your physician. The enterostomal therapist who worked with you at the hospital is also a good source of information, particularly if your questions involve the care of the ostomy. There are also excellent organizations that can provide both information and emotional support, such as the American Cancer Society and the United Ostomy Association. You can contact their main offices to find the chapter closest to your home.

RESOURCE LIST

If you have a problem with digestion or a question about your digestive system, the best place to go for information is to your own physician. He or she is in the best position to advise you about your problem. In addition, there are helpful special foundations and organizations for people with specific disorders and/or questions.

The American Digestive Disease Society, at 420 Lexington Avenue, New York City, is an excellent source of information on many digestive disorders.

The National Foundation for Ileitis and Colitis, Inc., 295 Madison Avenue, New York, NY 10017. Phone: (212) 685-3440.

The American Liver Foundation, 30 Sunrise Terrace, Cedar Grove, NJ 07009.

The United Ostomy Association, 2001 West Beverly Boulevard, Los Angeles, CA 90057. Phone: (213) 413-5510.

The American Cancer Society, 777 Third Avenue, New York, NY 10017.

International Association for Enterostomal Therapy, 505 N. Tustin Street, Santa Ana, CA 92705.

International Association for Medical Assistance to Travelers, 350 Fifth Avenue, New York, NY 10001.

The Food and Drug Administration, HFC 160, Rockville, MD 20857.

National Council Against Health Fraud, P.O. Box 1276, Loma Linda, CA 92354.

GLOSSARY

Abscess—a localized collection of pus surrounded and walled in by inflamed tissue

Acute—a symptom or disease of rapid onset, brief duration, and great severity

Arteriosclerosis—a condition of the arteries causing them to thicken

Bloating—a sensation of being filled with gas

Cholesterol—a fat-like material present in blood and tissues

Chronic—any condition continuing over a long period of time

Colostomy—a surgically created opening of the colon that drains outside the body through a cutaneous opening in the abdomen

Crohn's disease—a type of inflammatory bowel disease involving the colon (usually sparing the rectum) and/or the small intestine, usually localized to the ileum (ileitis)

Defecation—the act of moving one's bowels

Dehydration—water loss from tissues of the body

Diverticulosis—the presence of diverticula, small saclike openings, within the large intestine

Feces, fecal material—body wastes that are the end product of digestion

Fiber—the part of food that cannot be digested or absorbed and that causes stool to be bulky

Fistula—an abnormal communication between two or more contiguous organs

Flare-up—a sudden recurrence of symptoms

Hiatus hernia—a widened opening separating the esophagus from the stomach, sometimes associated with reflux esophagitis

Hormones—substances, such as estrogens and adrenalin, produced by specific glands. These substances enter the bloodstream and affect the body's functioning

Ileostomy—a surgically created opening of the ileum that drains outside the body through a cutaneous opening in the abdomen

Infection—an invasion of the body by harmful organisms creating a disease

Infectious gastroenteritis—a self-limiting intestinal infection characterized by nausea, vomiting, diarrhea, and fever

Inflammation—an abnormal condition resulting in swelling, redness, and heat

Inflammatory bowel disease—diseases of the intestine of unknown cause, characterized by recurrent bouts of bloody diarrhea; these diseases include ulcerative colitis and Crohn's disease

Irritable bowel syndrome—a functional disorder of the digestive tract manifested by assorted symptoms, including cramps, and diarrhea alternating with constipation

Meckel's diverticulum—a blind sac or pouch arising from the ileum

Mucus—a protective fluid produced by certain parts of the body (such as the surface lining of the digestive tract) to act as a barrier and a lubricant

Obstruction—blockage

Perforation—the creation of a hole in a tissue or organ

Peristalsis—involuntary wave-like motions, such as those that occur in the GI tract to help push food along the tract

Peritonitis—infection and inflammation of the inner lining of the abdominal cavity (peritoneum)

Protozoan—a parasitic, one-celled animal organism, such as an ameba

Reflux esophagitis—inflammation of the lining of the esophagus resulting from repeated regurgitation of irritating stomach contents into the esophagus

Regurgitation—the spilling back of material in the opposite direction from which it should normally flow

Stoma—an opening, usually surgically created, to allow for drainage

Traveler's diarrhea—a type of infectious gastroenteritis often occurring during travel outside one's own country and caused by the ingestion of contaminated food or drink

Ulcerative colitis—a type of inflammatory bowel disease involving the rectum and variable segments of the colon

INDEX